Practical Electricity and Electronics

THE MOTIVATE SERIES

Macmillan Texts for Industrial Vocational and Technical Education

Practical Electricity and Electronics

John Watson

MACMILLAN

Macmillan Education
Between Towns Road, Oxford OX4 3PP
A division of Macmillan Publishers Limited
Companies and representatives throughout the world

www.macmillan-africa.com

ISBN 0 333 60056 8

Text © John Richard Watson 1994
Design and illustration © Macmillan Education Ltd 1994

First published 1994

Cover Illustration courtesy of Science Photo Library/
Gregory MacNichol
Photographs Supplied by the author and Alan Thomas,
unless otherwise stated.

Printed and bound in Malaysia

2006 2005 2004 2003 2002
13 12 11 10 9 8 7 6

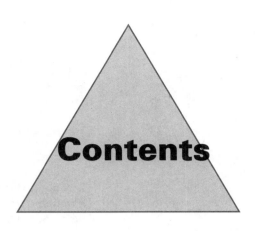

Contents

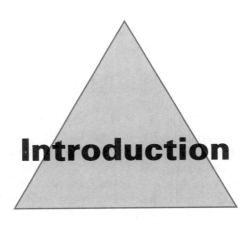

Introduction

The electronic age

Only 50 years ago, 'electronics' wouldn't have been listed in the courses offered by universities and colleges. It used to be no more than a specialist branch of electrical engineering, and was not thought important enough to be given a category of its own. Today we live in an 'Electronic Age'. Electronics is one of the most important of the technical subjects, and has probably changed the world and the way we live more than any other applied science. Society in countries all over the world has been affected.

Let's look at just a few of the most obvious examples of electronic equipment. Radio communications are now easy and relatively cheap. The first long-distance radio transmitter was designed by an Italian, Marconi, and in 1901 he sent Morse code transmissions – nothing as sophisticated as speech – across the Atlantic Ocean. The equipment Marconi used was huge (it filled a small building) and the transmitter used huge amounts of power. Yet today, just over 90 years on, we use portable radios that can send and receive clear speech or even colour pictures over hundreds of miles, and – via satellites – right round the world or even to the Moon.

Although radio was the first practical application of electronics, these days many aspects of the modern world depend on electronics: radio, telephones, television, facsimile machines, photocopiers, record and CD players, calculators, cameras, sound and video tape recorders, computers, radar, aircraft and ships' navigation and controls, medicine (from hearing aids to body-scanners), electric traction, production robots, missiles, music, even wrist-watches.

Think about that last example, the wrist-watch. Clockwork watches are expensive to make, need cleaning and oiling, and are not particularly accurate. The very best and most expensive wrist-watches (such as the famous Swiss-made Rolex watches) can achieve an accuracy of a few seconds per month, as long as they are looked after carefully and serviced regularly. The most famous Swiss-made *electronic* watch – the Swatch – costs about *one-hundredth* as much as a Rolex, and is more accurate. It doesn't need servicing, and when it goes wrong it is cheap enough to throw away. The Swatch is also assembled almost entirely by robots.

This shows another way in which electronics has affected the world: the way in which things are produced. For example, almost all cars made in the industrialised nations are at least partially assembled by robots – which are, of course, controlled by electronics. Today's cameras take far sharper pictures than the equivalent cameras of the pre-electronic age. Why? Because electronic computers are used to calculate the shape and density of the lenses. Previously, teams of mathematicians and craftsmen had to work for months or even years to design a good camera lens; but computers can do it thousands of times faster, cheaper, and better.

Alone among all the technologies, electronics has grown in importance and changed beyond all recognition in less than a single lifetime. To understand why this has happened, we need to look – briefly – at the history of the subject.

War in Europe

Three factors have been important in the rise of electronics. The first factor was World War II. This was perhaps the first war in which technology (rather than strategy and men) was important. The first large-scale use of aircraft – bombers and fighters – made it essential to improve radio communications and, in Britain, the invention of radar (**RA**dio **D**irection-finding **A**nd **R**anging) for locating enemy aircraft meant that there was an urgent need for more advanced electronic systems. Money and resources were devoted to research into the then relatively new science of electronics, both sides hoping that it would provide technical superiority and a strategic advantage. After the end of the war in 1945, the work carried out led to improvements in television, allowing the first commercial television service to begin.

Invention of the transistor

The second thing happened in 1957, when Bardeen, Brattain and Shockley, working at Bell Laboratories in the USA, assembled the first working transistor. This, more than anything, changed electronics completely.

In order to understand just how much impact the invention of the transistor was to have, you have to know that before the invention of the transistor every electronic machine required the use of **valves** (known as 'tubes' in some countries, including the USA). Valves will be described in more detail in Chapter 13, but for the time being it is enough to know that valves are to electrons what taps are to water; they control a flow of electricity.

Valves were invented in the early 1900s, and in the following 50 years had been improved in design but not in concept. Valves are quite large (typically the size of a man's finger), expensive to make, and extremely wasteful of power. Even the first transistors were quite small (about half a centimetre long) and used as little as *one-thousandth* as much power as valves would use to do the same job.

Quite quickly it became possible to mass-produce transistors on such a scale that the cost of each transistor was only a fraction of that of a valve. This made some electronic goods much cheaper (radio receivers, for example), but it also made possible the design and sale of very complicated electronic machines that would have been too expensive or completely impractical to make with valves (colour televisions and computers, for example).

The space race

The last step along the road to today's Electronic Age was the 'space race' in the 1960s. Russia and America competed to be first to land a man on the Moon, pouring billions of roubles and dollars into their projects. While Russia concentrated on building huge rockets with conventional but effective automatic control systems, the Americans chose to develop electronic and computer-controlled space vehicles in the expectation that they would be more adaptable. The Americans had less powerful rockets, which could not lift such heavy payloads, and so they had to build electronic systems that were very small and light. The result was that microelectronics and integrated circuits were developed as quickly as possible, more or less regardless of cost. It was money well spent, not just for America but for the rest of the world as well; everybody who uses a pocket calculator, portable television, or even a telephone, is reaping the benefits.

The point about transistors is that they don't have to be any particular size. A small transistor will work as well as a large one, and will use even less power. There is a lower limit, but that limit is unbelievably tiny. At first, transistor technology was used to make several transistors on the same piece of material (germanium or silicon), saving space and weight. Next, it became possible to make hundreds of transistors on a piece of silicon less than a centimetre square, and at the same time make the interconnections between the transistors to complete the circuit of, for example, an audio amplifier. A device like this is called an **integrated circuit**. Today, integrated circuits are made that include *hundreds of thousands* of transistors on a single chip of silicon – and improved mass production means that the price is still very low.

Here is a final example that will indicate just how far things have come in the last 50 years. One of the first working large-scale computers was made in England in the late 1940s. It occupied the whole of a large laboratory and used as much power as a small village. It was vastly expensive and – because it used valves – vastly unreliable. On average, one of its hundreds of valves had to be replaced every ten minutes. I now have a more powerful computer than that in my jacket pocket. It cost about as much as a cheap bicycle, and will run for 300 hours on two tiny batteries. It is so reliable that I would consider myself unlucky if it ever went wrong.

Milestones along the road to the Electronic Age	
1904	Fleming invents the electronic diode
1910	De Forest invents the triode valve
1957	Brattain, Bardeen and Shockley invent the transistor
1960–1969	The 'space race' leads to the development of integrated circuits

almost all countries have all helped to reduce the number of accidents.

If any area should be singled out as one in which there have been big advances in electrical – as opposed to electronic – technology, it is in **batteries**. A modern manganese–alkaline dry cell (for example, type MN1500) contains about *ten* times as much energy as its equivalent of 40 years ago. Rechargeable batteries of all types, from single cells to car and traction batteries, are much lighter, more efficient (less difference between the energy you put in and the energy you can take out), and have higher capacities.

Electricity

Although the spectacular advances have been in electronics, it would be wrong to underestimate the benefits that electricity has brought to communities everywhere. Electric power is now available in almost every corner of every nation, and has been responsible for great improvements in the living standards of millions of people. In most parts of the world reliable electric lighting is taken for granted; people are so used to it that they get quite angry if there are a lot of power cuts.

Electric power is easy to use, safe (if simple safety rules are followed properly), clean, and very flexible. It can be used for everything from lighting to railways. It can be generated from oil, gas, coal, falling water (hydroelectricity), atomic fission, waves, wind, or even sunlight.

There has been no 'revolution' in electric power, but there has been a steady improvement in the design, efficiency and reliability of electric appliances and machines. Electric motors, in particular, have improved almost beyond recognition in the last 50 years. Electric motors are more efficient (more power, less noise and heat) and – for a given power output – about half the size they were 30 years ago. The basic principles are still the same, but there have been advances both in materials and in design. And, of course, electronics are now used in motor and power control.

Electric lighting is also more efficient than it used to be (more light, less heat), thanks to advances in design and manufacturing technology.

Domestic and industrial electrical installations are now much safer than they were. Better design and materials, better working practices (through proper training), and strict safety regulations in

Electric power since 1940
• Steady improvements in appliances
• Improved electric power distribution
• Better safety
• Major improvements in motors and batteries

Learning about electronics

Electronics is quite a complicated subject, taking in some physics, some chemistry, some mechanical engineering, and even some drawing. There is a lot to learn, but the most important part of learning the subject is understanding the underlying principles. Once you have mastered these, learning the rest of it is relatively easy.

It is possible to put a lot of mathematics into electronics, but it isn't necessary to an understanding of the subject. Of course, mathematics is necessary when calculating what voltages, currents and components your circuits use, but the calculations are really very simple: more 'arithmetic' than 'mathematics'. In this book I have deliberately kept the mathematics to a minimum, just about the very least you need to get by. Your teacher will probably want to supplement what little I have included.

A study of electronics can be divided, broadly, into three rather separate disciplines: electricity, analogue electronics and digital electronics.

Electricity is basic to the whole thing; you need to know a certain amount about electricity before you can begin to understand anything about the way electronic devices work. And before you do much work with electricity, you must learn about

safety, which is where the main part of this book begins.

Analogue electronics (or **analog**, which is the American spelling) is the study of systems in which electrical quantities vary continuously. Examples are radio, record and tape players, and television receivers.

Digital electronics is the more recent aspect of the subject, and deals with electrical quantities that vary in discrete steps instead of smoothly. Many digital systems deal with only two possible values. The most important example is digital computers.

The layout and content of this book

As I mentioned above, I am beginning this book with a chapter on **safety**. Electricity can kill you if you're stupid or don't take the trouble to think, so make sure you read, understand and remember this before you go on.

Next I deal with **electricity** and **electric power**, which includes a summary of what you need to know about electricity in order to understand electronics. Your teacher may – or may not – want to supplement this material, depending on the details of the syllabus you are working to.

After a chapter on **measuring instruments** – essential if you are doing practical work – you will start 'electronics proper' by studying simple electronic devices (in this context, a 'device' is a single electronic component that would in practice be used in combination with others to produce a 'system' that will do something useful). Then we go on to look at various systems, putting together the devices and circuit to make a number of useful things.

> Many of the circuits and experiments given in the text can be used as classroom demonstrations, projects or for construction practice: all the circuits that are suitable for building are marked with a '★' in the caption.

The first systems to consider are **amplifiers** and **oscillators**. Then we shall look at **radio and television**, the aspect of electronics that has probably done the most to change the way we live and what we know about the rest of the world.

Because electronics is essentially a *practical* subject, there is a chapter that talks about the way commercially made electronic systems are assembled, how you can test them, and how (sometimes!) you can repair them.

The final part of *Practical Electricity and Electronics* begins your study of **digital electronics**, if the syllabus you are studying calls for it. Even if it doesn't, I am sure you'll find it interesting to read if you have time. Although – for example – a computer is much too complicated to look at so early in your studies, you will at least learn about the basics of **digital logic**, which is as much the foundation of computer science as electricity is essential to understanding electronics.

I hope you will enjoy this book. I have certainly enjoyed studying and working with electricity and electronics, largely because the two technologies are so much part of the modern world, and have done so much to improve people's lives. I have tried not to be too formal, and that is because there is no reason to make a serious subject like electronics dull. Science and technology ought to be *fun* as well as interesting, productive and vitally important for the future of mankind.

Construction notes

Many of the circuits shown in this book – those marked '★' in the captions – are suitable for construction by students. I have tested all of them, and have used most of them in the classroom. Most of them can be built using 0.1 inch matrix Veroboard or some similar printed-circuit strip-board. The larger 0.15 and 0.2 inch matrix boards may be better if you are inexperienced with the use of the soldering iron. A better alternative, especially if the components have to be re-used, is a plug-in prototyping 'plug-board', preferably with 0.1 inch spacing.

None of the circuits is especially critical about layout. I have not, therefore, provided any layout diagrams. Students will benefit from working from the circuit diagrams, and if you start with the simpler designs, you will be able to progress steadily to the more complex ones without problems. Only one or two circuits require high voltages, and the usual precautions should be taken: isolated power supplies, and careful supervision.

Electrical safety

Introduction

In learning about electricity and electronics you will be working with electric power, and that can be potentially dangerous. That is why this book begins with a chapter on safety. Electricity is really quite a safe form of power if used carefully. It can't blow up like oil or petrol, and it won't suffocate you like gas or fumes from coal. It isn't poisonous. But it can start fires, and it can kill you by electric shock.

Safety is important! Electrical and electronic engineering are not particularly dangerous occupations, but it is essential to know basic safety rules and procedures, to be aware when you are getting into a situation that might be hazardous, and to know what to do if the worst happens.

The most common hazards are electric shock, fire and accident (meaning accidents that involve tools and machinery).

Electric shock

Most, but not all, injuries and deaths from electric shock are caused by contact with the mains electricity supply. In most countries the mains electricity supply is 220–240 volts a.c. (alternating current), although some use 110 volts. Factories often use higher voltages (such as 440 volts), and the electricity distribution industry uses very high voltages in substations and the like.

A domestic electricity supply involves three wires: the **live** wire, the **neutral** wire and the **earth** wire. Only one of these, the live wire, is dangerous in normal circumstances. It is 'live' with respect to earth. That means that 240 volts (or whatever the mains supply voltage might be) will flow from the live wire through any conducting material or object that is connected to, or standing on, the ground. If that object happens to be a person, the result can be serious injury or even death.

> ▲ The most important two rules for protection against electric shock are these:
> - If a wire or piece of equipment is live, or you think there is the slightest chance that it might be live, don't touch it.
> - Always think before you touch any wiring.

Electric shock kills by temporarily paralysing the heart muscles, so the greatest danger occurs when the electric current flows across the chest. This is most likely to be from one arm to the other, or through an arm down through the legs.

The electrical resistance of the human body depends mostly on the resistance of the skin. If the skin is damp, either from water or because the weather is hot or humid, the resistance is dramatically decreased, and the chances of a shock being fatal are greatly increased. So you should never operate electrical equipment with wet hands.

Although the most serious shocks usually come from the mains supply, remember that battery-powered portable equipment can sometimes generate high voltages that are dangerous. Electronic flashguns and portable televisions are examples. It pays to be careful. If in doubt, don't touch.

What to do for a victim of electric shock

Get the victim away from the electricity
Get the victim away from contact with the elec-

tricity. The best way to do this is to switch off the current. Don't touch the victim until you are sure that the power is off, or you could become a second victim yourself!

If you cannot switch off the current, drag or push the victim away from contact with the electricity supply by using a suitable insulator (Figure 1.1). A wooden or plastic chair, a piece of dry wood or a broom handle would do, or you could hook a rope or piece of insulated cable around the victim's arm or leg.

Figure 1.1 Using a non-conductor to move a person away from a live wire.

If someone is in contact with a high-voltage mains supply, such as an electricity company sub-station transformer, do not attempt a rescue unless you are certain the power is turned off. The victim is probably dead anyway, and a wooden pole would not be a sufficiently good insulator at high voltages to prevent the would-be rescuer from being killed as well. Get expert help!

Has the victim's heart stopped beating?

Once the victim is away from the electricity supply, you can think about what to do next.

If he is unconscious and seems not to be breathing, try to find out if his heart has stopped. Blueish lips, enlarged pupils (lift an eyelid back carefully to check the pupils), and no pulse are signs that the heart has stopped. Pull open the victim's clothing and listen for a heartbeat by putting your ear against his chest, just below and to the left of the breastbone. Feel for a pulse in the artery at either side of the neck.

If you are sure that there is no heartbeat – *and only if you are certain* – someone should attempt external cardiac massage to restart the heart. This is a job for an expert, so try to get someone who has been trained.

Try cardiac massage yourself *only if a trained person is not available within a minute or two.*

If you are sure the victim's heart has stopped,

and no trained help is available, use this procedure for external cardiac massage, to try to restart the heart. If the heart has not stopped, you could do more harm than good, so be certain.

1. Lie the victim on his back on the floor.
2. Apply one breath of artificial respiration, as described below.
3. Find the right place, about one hand-span down from the top of the collar-bone and just to the left of the breastbone, and thump the victim's chest once with your clenched fist, bouncing your fists off the ribs. If you are lucky, this will start the heart beating, so check for a pulse and listen for the heartbeat.
4. If the heart is still not beating, place both hands on the lower end of the breastbone, straighten your arms, and press down rapidly, hard enough to push the breastbone down about 4 centimetres (1.5 inches). You will need to press quite hard, but not so hard that you break any bones.
5. Press down rapidly once per second, just five times.
6. Then apply one breath of artificial respiration. Check for a heartbeat, and if there is none, repeat the procedure, giving another five compressions and another breath before checking again for a heartbeat.
7. Immediately you detect a heartbeat, stop external cardiac massage, but carry on with artificial respiration until the victim is breathing on his own.
8. Get trained medical help as quickly as possible.

Has the victim stopped breathing?

If you can detect a heartbeat, check to see whether or not the victim is breathing. If you don't think he is, then someone should attempt artificial respiration. If someone who has been trained is available, get them to do it. Otherwise do it yourself. The easiest method is mouth-to-mouth.

Lie the victim on his back. Tip his head back, so that you are looking down his nostrils, and pull his jaw forward to move his tongue away from the airway. (Sometimes this starts the victim breathing again on its own.) Pinch the victim's nostrils together, so that the air you are going to give him won't simply come straight out again.

Take a deep breath, and put your mouth over the victim's lips, making as airtight a seal as you can (Figure 1.2). Blow into his lungs four times, blowing from your chest (not from your cheeks). The victim's chest should rise visibly each time you blow. Blow just hard enough to inflate the chest.

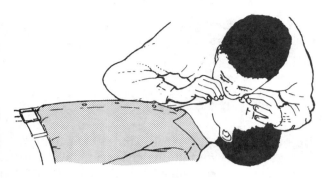

Figure 1.2 Mouth-to-mouth resuscitation.

Keep on trying!
Carry on with this procedure, trying to keep up a steady pace – say 10 to 15 breaths a minute – until trained help arrives. People have been known to survive after several hours of artificial respiration, so don't give up.

The recovery position
If the victim is unconscious but breathing on his own, and if you are sure that he has not sustained any serious injuries other than the electric shock, you should put him into the recovery position. The recovery position enables the victim to breathe easily, and reduces the chances of his choking if he vomits.

Roll the victim onto his side, and move the uppermost arm so that it makes a right-angle with his body. Move the uppermost leg so that it makes a right-angle with the body, too. Turn his head from side to side, and tilt it back, with the chin pushed forwards, so that the airway is clear. Figure 1.3 shows how the recovery position should look.

Figure 1.3 The recovery position.

Electrical burns
Be especially careful about any burns. Electricity can cause extremely serious burns, which may not look very large on the surface, but which can be very deep. Even if the victim seems to be recovering, get trained medical help.

This is not a book on first aid, and the details given above are by no means exhaustive. If you can get one, read a book on first aid. You could save someone's life.

Electrical safety

Sources of high voltage are clearly dangerous, but in some cases sources of high **current** are dangerous as well.

A car battery is the most common source of dangerous current. You could safely grasp the terminals of a normal 12 volt car bettery with your hands and not even feel it because the voltage is quite low. But, if the terminals of a car battery are connected together by something that is a good conductor of electricity – metal – then the result can be very hazardous.

The very large current that flows can heat up a metal conductor – which might be a bracelet, necklace, or even a metal watch-strap – so quickly that it becomes red-hot in a second. The current can be so strong that it burns right through the metal. Either way, serious burns can be caused this way. Connecting the terminals of a bar battery together with a very thick piece of metal – such as a spanner – can make the battery explode, causing serious injuries or even death.

> ▲ Avoiding electrical accidents is mostly a matter of applying common sense, and making sure that equipment is properly connected and used.

There are special precautions to be taken when you are working on electrical installations, as safety is one of the first considerations in any electrical installation.

In the home, laboratory, or workshop, remember that apparatus designed to be connected to the mains should be earthed, unless it is clearly marked as being double insulated. Figure 1.4 compares the way in which the two methods of protection work for faulty equipment, showing how the user is protected. Earthing or double insulation makes electrical equipment relatively safe to use, but *not* if you take the covers off for working on the interior of the equipment.

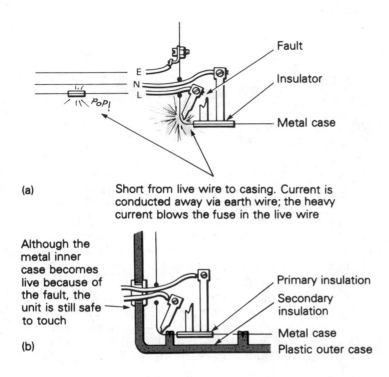

(a)

Short from live wire to casing. Current is conducted away via earth wire; the heavy current blows the fuse in the live wire

Although the metal inner case becomes live because of the fault, the unit is still safe to touch

(b)

Figure 1.4 (a) An earthed appliance; (b) a double-insulated appliance.

An **isolation transformer** provides a measure of safety when working on equipment that requires main voltage. An isolation transformer is a transformer (see Chapter 2 for the way a transformer works) that has two separate two windings, insulated from each other to a high standard. The output voltage is the same as the input voltage but safety is improved because the output is isolated from the mains supply, and neither terminal is at a high voltage with respect to earth. This means that touching just one connection or the other will not result in a shock. The voltage is still just as high, however, and simultaneous contact with *both* wires can be lethal, so all the usual precautions need to be taken.

Take extra care when switching off circuits to isolate them from the mains supply. Anyone can make mistakes. I am lucky to be writing this book, for a few years ago I was working on a mains line. The line was controlled by a double-pole isolating switch of approved design. Before starting, I checked that the switch was off, double-checked in fact. I then took my cutters and carefully cut the

wire on the wrong side of the switch. It cost me a good pair of cutters, but I could have been killed.

Fire

Electrical equipment of all kinds dissipates heat, and under fault conditions the heat output of a component can rise high enough to start a fire. Even everyday items of equipment can be a hazard. I once saw a colour television with most of its insides and part of a plastic cabinet completely devastated by a fire caused by a faulty component.

The component overheated and melted a plastic insulating sleeve over high-voltage connections to the back of the tube. This in turn caused a massive high-voltage short circuit which set fire to the whole receiver. The owner was fortunate that the fire confined itself to the television.

 Fires can start unexpectedly. Be on the alert!

When to tackle a fire

It is important to realise that you must only tackle a fire: (1) if you think it is small enough to put out – don't overestimate the power of even large extinguishers; (2) if there is no personal danger; (3) *after* you have raised the alarm. The first rule of dealing with an electrical fire is turn the current off first. If you can't pull out the mains plug or get at the switch on the equipment, turn off the power at the mains switch.

Fires can be put out by removing the combustible material (not usually possible), by removing the oxygen supply, or by cooling the burning material below the point at which combustion can be sustained.

Fire extinguishers

Various things can be used to put out a fire.

Water
Depending on what is on fire, water might be appropriate for putting it out, but never, never use

water if there is the slightest chance that the power might still be on. Water must never be used on burning liquid, as it will just spread the liquid and the fire.

Sand or dry soil
This works well on small fires, but should not be used on burning liquids, as it will just spread them around.

Foam extinguishers
These work by smothering the fire in a foam containing the inert gas carbon dioxide, and hence depriving it of oxygen. However, foam extinguishers contain water and are not suitable for electrical fires.

Carbon dioxide extinguishers
These contain liquid carbon dioxide gas, which smothers the fire and starves it of oxygen. Carbon dioxide extinguishers are suitable for all types of small fire, including burning liquids. Although the gas is very cold when it comes out of the nozzle, it has little cooling effect on the fire so the fire might re-ignite if the gas is blown away. If you are using a carbon dioxide extinguisher, don't hold the flared nozzle in your hand; it gets so cold that it can 'burn' you.

Dry powder extinguishers
These are useful for electrical fires, especially fires that are fairly confined (such as a television or item of test equipment that has caught fire). The dry powder does no damage and can be brushed off afterwards, so there is more chance of salvaging apparatus that has been extinguished with dry powder. Dry powder extinguishers are not suitable for large fires.

Fire blankets
These can be very effective for small fires. Fire blankets are usually made of aluminised fibreglass cloth, and are simply placed over the burning object to deprive it of oxygen. This means going right up to the fire, so be careful. In an emergency, if a proper fire blanket is not available, a wet towel, wet blanket, or even a wet coat can be used instead.

If you can't put the fire out yourself . . .

If an item of equipment, furniture, or even a whole room is ablaze, and you decide not to tackle the fire yourself, *shut the door as you leave to raise the alarm*. This will slow the fire's spread, even if the door is only an ordinary one. It will also slow the spread of smoke and toxic fumes, which are potentially more deadly than the fire itself.

- • If there is a fire alarm, sound it immediately.
- • Get everybody out of the building.
- • Call professional help to put out the fire.

Fire safety

As a preventive measure, don't leave equipment running if there is nobody about. Electric heaters are frequent causes of fires.

Make sure all fire doors, fire escape routes and fire appliances are in usable condition.

If there is a fire alarm system, make sure it is tested regularly.

If you are in an office, factory or public building, there should be regular fire drills.

CHECK YOUR UNDERSTANDING

● Get a victim of electric shock away from the electricity as quickly as you can – either turn the power off or move him with a non-conductive pole or something similar.

● Get trained help! Only try to help him yourself if none is available.

● Electrical burns may not look bad, but can be serious; get them treated.

● Be aware of the dangers of electricity. High voltages can cause death by **electric shock**, and sources of high current – especially car batteries – can cause serious **burns** if the terminals are connected by anything that conducts electric current.

● If there is a fire, raise the alarm first!

● Tackle only small fires yourself.

REVISION EXERCISES AND QUESTIONS

1 You see someone lying on the ground. He has been working on some cables and you think he may have been hurt by electric shock. What is the **first thing** you do?

2 If you have to give someone mouth-to-mouth resuscitation, for how long should you keep trying if there are no signs of life?

3 If someone has a burn caused by an electric shock, is it likely to be

i) very serious;
ii) painful but not serious; or
iii) not at all serious?

4 A television set you are working on suddenly bursts into flames. Before you can do anything about it the plastic cabinet is burning fiercely. What should you do next?

5 A car battery is completely safe because, being low voltage, it cannot give you an electric shock. Is this true?

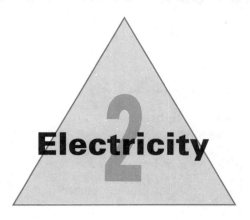

Electricity

Introduction

We usually recognise electricity by the way it changes materials and objects in the world around us. Electricity is, after all, invisible and silent. Before you can begin to study electronics, you need to know about electricity. And in studying electricity, we must begin by considering what it does to other things.

Electrical effects

You cannot see, hear, or even measure electricity directly, but it can be detected by the way it affects matter. Electricity interacts in five important ways: to produce **heat**, to produce **light**, to affect **chemical reactions**, to cause and be caused by **magnetism**, and to have a **mechanical** effect. This section outlines these important effects of electricity, and where possible describes simple demonstrations to illustrate the points.

Heating effect

One of the most important physical effects caused by electric current is **heat**. The heating effect of an electric current is used in many items of electrical equipment – cookers, electric heaters and lamps, for example – and it is something that must be considered when designing any electrical apparatus.

Without considering how it works at this stage, let us consider a source of electrical energy. It might be a battery, or a special laboratory power pack. If this is connected to an electric lamp, as shown in Figure 2.1, electrical current will flow through the lamp.

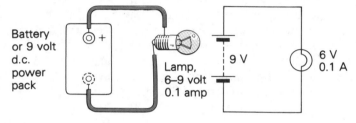

Figure 2.1 *A simple circuit with circuit diagram.

After a minute or so, carefully feel the envelope of the lamp. It will be warm, perhaps even hot. If it is not too bright, look closely at the filament inside the lamp. What makes it light up?

The answer is that the filament is glowing white hot. The envelope of the lamp is filled with an inert gas, usually argon, to prevent the filament from burning up in an instant.

The lamp lights only when both wires are connected to the power supply. Electric current must be able to flow from the power supply, through the lamp, and back to the power supply. As the electric current flows round in a circle, we call this an electrical **circuit**.

Figure 2.2 shows a battery (or power pack) converting electricity into heat. Heat is dissipated by a component called a **resistor**; see below for an explanation of this component.

A power pack or battery capable of providing at least 0.5 amps should be used. The resistor will get hot to the touch after a short time.

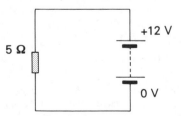

Figure 2.2 *Power: a simple circuit like this will dissipate heat.

Lighting effect

Electric current can be used to produce **light**. In the section above, we saw how a lamp filament can be heated to such a high temperature that it glows white hot and gives off light. But this is not direct conversion of electrical energy into light. Only recently has 'light without heat' become a reality. It can be demonstrated very easily using a light-emitting diode or LED.

Figure 2.3 shows a circuit with suitable values for the components. When the power supply is connected the LED will light up (the LED has to be connected the right way round). The light produced by the LED is a result of electrical energy being converted directly into light. Nothing has to heat up to produce the light. The mechanism by which the light is produced is quite subtle, and is dealt with in Chapter 9 of this book.

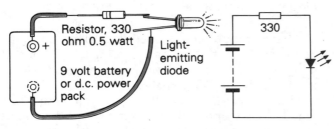

Figure 2.3 *A light-emitting diode (LED) in a simple circuit.

Chemical effects

Electrolysis

The chemical effect of an electric current was one of the first electrical effects to be discovered. It is very important in all sorts of applications.

Anyone who has studied physics will know that matter consists of atoms of different elements and that the elements are bound together in molecules to form **compounds**.

Some examples of elements are oxygen (symbol O), copper (symbol Cu), sulphur (symbol S), and hydrogen (symbol H). Examples of compounds formed from these elements are water (symbol H_2O – two hydrogen atoms plus one oxygen atom), sulphuric acid (symbol H_2SO_4 – two hydrogen atoms, plus one sulphur atom, plus four oxygen atoms), and copper sulphate (symbol $CuSO_4$ – one copper atom, plus one sulphur atom, plus four oxygen atoms).

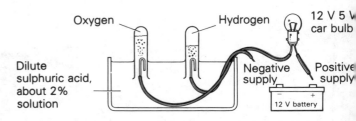

Figure 2.4 *Electrolysis: the chemical effect of an electric current. To fill the tubes, immerse them in the bowl, then carefully move them upright. They can be supported with a retort stand.

One of the best demonstrations of the chemical effect of an electric current is electrolysis. If an electric current is passed through water (H_2O), some of the molecules will be decomposed to form oxygen (O_2) and hydrogen (H_2) gases. The apparatus shown in Figure 2.4 can be used for this demonstration.

For safety's sake, it is essential that you use only a battery power supply. Do *not* use a power supply connected to the mains. A 12 volt car battery is ideal, but when using car batteries it is essential to make quite sure that wires connected to the two terminals never come into contact, as the heating effect of the electric current can cause a fire or even an explosion if this should happen. The car bulb included in the circuit will not normally light up, but is a safety feature to limit the amount of current that can flow: see below.

> ⚠ This demonstration must be carried out in a place that is well ventilated. Gases are given off that can explode. Keep any flame well away from the equipment.

With pure water, the experiment will work rather poorly, because pure water is not a good conductor of electricity. Add a little dilute sulphuric acid (10% solution is available at chemist shops; add a little of this to the water) to improve the water's conductivity. If dilute sulphuric acid is not available, some ordinary salt, added to the water, can be used instead.

With the apparatus set up as shown in Figure 2.4, gas will start to bubble from the electrodes, showing that the water is being dissociated into its component gases by the electric current.

Electroplating

The same simple apparatus can be used to demonstrate another useful chemical effect of an electric current: electroplating. This is the process used in making chromium-plated articles.

Set up the apparatus, but fill the bowl with a strong solution of copper sulphate ($CuSO_4$) dissolved in water (*Caution: copper sulphate is poisonous*). The end of the lead connected to the **positive** terminal of the battery, known as the **anode**, should be made of copper. A copper nail, a piece of copper sheet, or even a spiral of copper wire will work. Twist the connecting wire firmly round the copper to ensure a good connection. The end of the lead connected to the **negative** terminal of the battery, known as the **cathode**, should be made of iron or tinplate. The end of a tin can is quite suitable. Make sure it is completely clean.

When you connect the battery, watch the iron (or tinplate) cathode. After a short time it will discolour. Leave it for a few minutes, then disconnect the battery, remove the iron cathode and wash it. If you polish it, you will see that it has been copper plated! Copper has been dissolved from the anode, carried through the solution, and deposited on the cathode. This is a convincing and useful illustration of the chemical effect of an electrical current.

Magnetic effect

Perhaps the most widely exploited effect is the magnetic effect of an electric current. This is used in all sorts of devices, from electric motors to speakers. Whenever an electric current flows along a conductor, such as a wire, it is accompanied by a magnetic field. The simple apparatus shown in Figure 2.5 can demonstrate this. When the leads are connected to the battery (or power pack) a current flows through the circuit. The bulb is included in the circuit for two reasons: first, to indicate that the current is flowing; second, to limit the current to a safe level. Figure 2.6 shows a closer view of the compass.

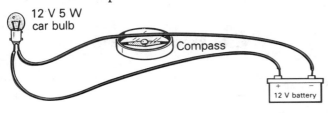

Figure 2.5 *Magnetism: the magnetic effect of an electric current.

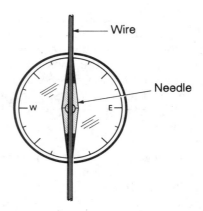

Figure 2.6 A close-up view of the wire and compass alignment.

Make sure that there are no large metal objects close to the compass. With the current off, the compass needle should align itself along a north–south axis. Carefully place the wire so that it lies exactly along the line of the needle; electrical wires are made of non-magnetic metals, so the needle's position will not be affected. Now connect the current. The lamp will light, and the needle will be deflected. Without disturbing the compass or the wire placed across it, reverse the battery connections. How does the behaviour of the needle differ from the way it behaved in the first demonstration?

It is clear from this simple demonstration that where there is a flow of electric current, there is also magnetism.

Mechanical effect

It is possible for electrical energy to be turned directly into movement. This is the mechanical effect of an electric current. It is more difficult to demonstrate with simple apparatus than the other effects, as the amounts of movement produced are quite small.

Crystals of certain materials, such as quartz, Rochelle salt (sodium potassium tartrate) and some ceramics, have the property of expanding along one axis and contracting along another when subjected to an electric field. This is known as the **piezoelectric effect**.

Such crystals or ceramics are used in the sounders ('bleepers') that are built into digital alarm watches. You can see the sounder if you remove the back of a digital alarm watch (don't use a new one!) and examine the insides. The sounder is usually mounted on the inside of the back cover. It looks like a very thin circle or square of ceramic material. You will be able to find a contact spring, projecting from the watch movement, that touches the sounder when the watch is closed. The case forms the second terminal.

When the alarm goes off, the watch circuits apply an electric voltage to the sounder, turning it on and off very rapidly, at the rate of about 1000 to 2000 times per second. As the voltage is applied, the sounder bends very slightly. When the voltage is turned off, it relaxes. This rapid movement produces vibrations in the air and in the back of the watch case, which are audible as a 'bleep' sound. Figure 2.7 shows how this works, exaggerating the movements of the sounder, which are actually very tiny and too small to see with the naked eye.

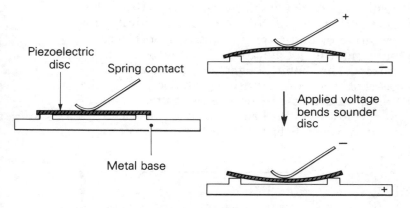

Figure 2.7 The piezoelectric effect.

The piezoelectric effect is also used in the pick-up cartridge of cheap record-players. Sound is recorded on the record in the form of a wavy groove, along which the sapphire or diamond pick-up stylus moves. As it tracks along the groove, the stylus is vibrated from side to side by the groove, to reproduce the recorded sound. The tiny movements of the stylus are converted into a varying electrical voltage by a piezoelectric crystal which, when it is bent by the movement of the stylus, generates electricity. Figure 2.8 shows this diagrammatically.

Have you noticed that this is the reverse of the piezoelectric effect of the sounder? A varying electric voltage was applied to the sounder to make it bend. In the record-player pick-up, the piezoelectric crystal is bent by the stylus, and responds by producing an electric voltage.

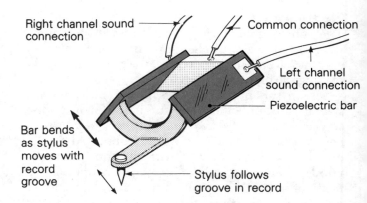

Figure 2.8 The pick-up cartridge of a record-player.

Some good-quality cigarette lighters make use of the piezoelectric effect to make a spark. A stack consisting of a very large number of piezoelectric crystals is struck by a metal hammer when you press the button on the lighter. This bends all the crystals in the stack. It bends them only very slightly, but enough to produce quite a high voltage. The high voltage makes a spark jump in front of the gas jet, igniting the gas.

The effects of an electric current are reversible

In practice, all the electrical effects demonstrated above can be reversed.

1. It is possible to produce electric current directly from heat.
2. It is possible to use light to generate electricity: solar cells are quite commonly used to power electrical equipment, from watches to spacecraft.
3. Chemical changes can generate electricity. Batteries do this.
4. In generating stations, magnetism is used to produce the mains electricity supply.

The heating, chemical and magnetic effects of electricity are very important, and each topic is studied in detail in the relevant parts of this book.

Electric circuits

Electricity

What is electricity? This seems to be a good question with which to begin.

To answer this question, we have to look at the composition of matter. All matter is, as we saw earlier, made up of atoms, but it is far from simple to describe an individual atom. From a study of physics, any student will know that atoms are made up of a **nucleus**, around which orbit **electrons**. But nobody really knows what an atom looks like, as the largest atom is far too small to see even with the most powerful microscope. So physicists design models of atoms to help them to explain atomic behaviour.

One of the simplest and most straightforward models of the atom was proposed by Niels Bohr, a Danish physicist, in 1913. It is Bohr's model that we most often think of, with its tiny electrons in orbit round the heavy nucleus. A Bohr atom is illustrated in Figure 2.9.

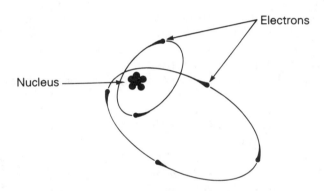

Figure 2.9 A model of an atom, according to Bohr's theory.

The electrons are confined to orbits at fixed distances from the nucleus, each orbit corresponding to a specific amount of energy possessed by the electrons in it. If an electron gains or loses the right amount of energy, it can jump to the next orbit away from the nucleus, or towards it. Electrons in the outermost orbits are held to the nucleus rather more weakly than those nearer the middle, and can under certain circumstances be detached from the atom. Once detached, such electrons are called **free electrons**.

It is important to realise that the gain or loss of electrons does not in any way change the substance of the atom. The nucleus is unchanged, with the same number and kind of particles in it, and so an atom of, say, copper can lose or gain electrons and still remain copper.

Each electron carries one unit of negative electric charge. In a 'normal' atom, the charges on the electrons are exactly balanced by the charge on the nucleus. An atom of copper normally has 29

electrons in orbit around the nucleus. Each electron has one unit of negative electric charge, and the nucleus has a total of 29 units of positive charge.

If a copper atom loses an electron, the nucleus will be unchanged. But as it still has a total of 29 units of positive charge, and there are only 28 electrons and thus 28 units of negative charge, the atom has overall, one unit of positive charge that is not balanced by a corresponding negative charge. Such an atom is called a positive ion, sometimes called a **cation** (pronounced 'cat-iron').

Similarly, atoms can gain extra electrons. If a free electron meets a neutral atom, the electron may go into the outer orbit around the nucleus. In this case, there will be one more negative charge than is needed for neutrality. Such an atom is called a negative ion, sometimes called an **anion** ('an-iron').

What is an electric charge? My technical dictionary says it is '... *the quantity of electricity in a body*'. When I look up 'electricity', it says: '*The manifestation of a form of energy believed to be due to the separation or movement of . . . electrons.*' I don't think this is very helpful as an explanation!

The real answer is that it is impossible to say in words just what electric charge is. It can be described mathematically, but this is not the same as describing it physically. At least we have a very clear and detailed idea of how electricity behaves, and this enables us to use it in all sorts of clever ways without actually needing an underlying understanding of the nature of electricity and electric charge.

When looking at the physics of electricity, it is wise to remember that we are looking at models rather than the real thing. I shall try to remind readers of this book about this once in a while.

Circuits and circuit diagrams

A good place to begin is to take a simple electrical circuit, and then look at its constituent parts.

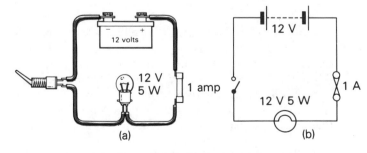

Figure 2.10 *A simple circuit shown (a) as a picture and (b) as a circuit diagram.

Figure 2.10 shows a simple battery and lamp circuit in two forms. In Figure 2.10(a) it is shown as a picture, and in Figure 2.10(b) it is shown as a circuit diagram.

Like most circuits, this circuit can be divided into basic parts: a source of energy, conductors of electricity, a load, a control, and protection.

E.m.f. and p.d.

In this circuit the source of energy is a battery. In Chapter 3 we shall look at batteries in more detail, but for now you need know only that a battery is a source of electric power. We could use a different source: two other sources of electrical energy are generators or solar cells. The important feature is the fact that between the terminals of the battery there exists a **potential difference** (abbreviation p.d.). Potential difference is measured in **volts** (symbol V), named after the Italian physicist Alessandro Volta, who made the first practical battery. A potential difference is simply a difference in total electrical charge. Electrochemical reactions in the battery cause one terminal to contain many positive ions, and the other to contain many negative ions.

The battery is a source of **electromotive force** (abbreviation e.m.f.). Like p.d. it is measured in volts. There is a subtle distinction between the two. E.m.f. is the p.d. of a source of electrical energy, such as a battery. P.d., however, measures the difference in electrical potential between any two points regardless of whether or not they are a source of energy; for example, it is possible to measure the p.d. across the terminals of an electric lamp, but nobody would suggest that the lamp is a source of electrical energy!

The e.m.f. of the battery in Figure 2.10 is 12 volts.

Conductors and cables

A conductor is a material that electric current will flow through quite easily. All metals are good conductors, as are some other materials such as carbon. Almost all plastics are very poor conductors – insulators, in fact. An insulator is simply a material that is a bad conductor of electricity. One of the best insulators is glass. Most ceramics are insulators, along with rubber, oil and wax.

A flow of electric current through a conductor consists of free electrons moving from atom to

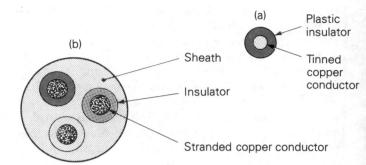

Figure 2.11 A cross-section through two types of cable.

atom through the material. In order for there to be a useful amount of current, a very large number of electrons must flow. Accordingly, the basic unit of electric current flow is equivalent to around 6.28×10^{18} electrons per second moving past a given point in the conductor. This unit of electric current is called the **ampere** (often abbreviated to 'amp'), and is named after André Marie Ampère, a French physicist who did important work on electricity and electromagnetism. The ampere is given the symbol A.

We are all familiar with electrical wires. Figure 2.11 illustrates a cross-section through two different types. Look first at Figure 2.11(a). The wire consists of a central conductor that is surrounded by flexible plastic insulation. The conductor is most likely to be made of tin-plated copper. Copper is one of the best conductors of electricity (only silver is better) and is also fairly flexible and relatively cheap. It is tin-plated to prevent the surface of the copper from oxidising; copper oxide is a poor conductor of electricity, which could give trouble if you used the cable with a screw-type connector. An oxidised surface is also difficult to solder.

The insulation surrounding the conductor is usually made of polyvinyl chloride (PVC), a flexible plastic with excellent insulating properties and an extremely long life in normal use. In the past, rubber insulation was used, but rubber eventually perishes and cracks.

For low-voltage circuits like the one in Figure 2.10 it is unnecessary to have insulation that will withstand high voltages. You could hold on to the 12 volt battery terminals quite happily without getting an electric shock. This is not the case with high-voltage circuits (such as the mains supply), where insulation is vitally important to a safe installation.

Figure 2.11(b) shows a typical cable used in house wiring. This cable has three cores, or conductors. There are two insulated conductors, with the insulation coloured red and black to indicate which is which. A third (earth) conductor is also present, but this does not have any insulation as it is not meant to carry current under normal circumstances. Mains three-core electricity cables always have the core colour-coded so you can tell which are which.

The **live** (dangerous!) core is coded **red**, or in some countries (including Europe), **brown**. The **neutral** core is coded either **black** or **blue**. The **earth** is either uninsulated or coded **green**; sometimes it has **green and yellow stripes**. A second layer of insulation, the **sheath**, covers the three cores. The current-carrying conductors are thus insulated by two layers of PVC. The sheath not only provides insulation, but also gives mechanical protection to the insulated cores inside.

The types of cable shown in Figure 2.11 will bend, although they are quite stiff. If a more flexible cable is needed, the cable to an electrical appliance for example, the conductor is manufactured with several strands of thinner wire. This allows the cable to bend more easily and to be bent more often without danger of the cores fracturing.

Cables come in all sorts of different types. The photograph in Figure 2.12 shows a selection of cables, ranging from simple insulated wires to multicored 'ribbon cable' used for carrying data signals.

Loads

The load in the circuit in Figure 2.10 is a lamp. In an electric circuit, a load is any device that uses electric power and dissipates energy. The lamp converts electrical energy into heat (and some light); this energy, which comes in the first place from the battery, leaves the circuit entirely.

A 'load' in an electrical circuit can be one of a large range of devices: everything from an electric motor, lamp or bell to a house, which might be the load in a circuit of a large generator, or even a whole city.

Control

Almost every electrical device that uses power needs some form of control. Usually this takes the form of a switch or circuit-breaker, to interrupt the flow of electric current in the circuit.

Switches can be almost any size, according to the work they have to do. The circuit in Figure

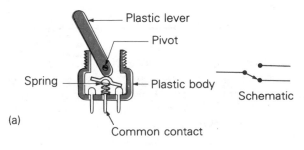

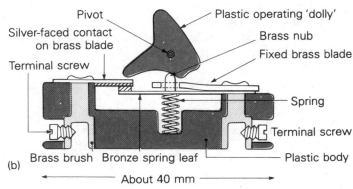

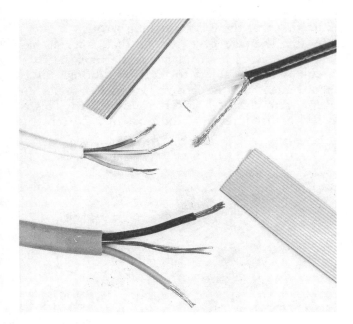

Figure 2.12 A selection of common cables and wires.

Figure 2.13 (a) A low-voltage change-over switch; (b) a modern slow-break microgap mains switch.

2.10 requires only a small switch, as the amounts of current and voltage involved are quite small. Circuit-breakers used at power-generating stations have to interrupt very large currents and voltages, and are sometimes the size of a small house. Figure 2.13 shows the mechanism of two typical switches.

Figure 2.13(a) illustrates a low-voltage 'change-over' switch, designed for use in low-voltage electrical equipment. Switches are rated according to their working voltage and current. This type of switch could interrupt currents up to about 1 amp at voltages up to 100 volts; it is useful for battery-operated appliances, but unsuitable for mains applications.

Figure 2.13(b) shows a typical switch used in a house for controlling the lights in a room. It is intended for use at voltages up to 250 volts, and currents up to 3 amps.

Protection

> ⚠ A car battery is dangerous. It gives off hydrogen, a flammable gas that can explode in air, when it is being charged. Always charge car batteries in a well-ventilated place.

A car battery is also a source of considerable energy. Although the voltage is low enough to avoid risk of shock, the amount of current that a car battery can deliver is substantial. If you were to connect the two terminals of a car battery together with a wire, the wire would immediately melt or burn. If you used a very heavy wire, the battery could explode. Either way, you would be in danger of serious injury.

In the circuit of Figure 2.10 it should not be possible for this to happen. But accidents can always occur, so in any electrical system that has the potential for dangerous currents or voltages, protection devices are used.

The simplest protection device is a **fuse**. A fuse simply consists of a thin wire, often sealed in a glass or ceramic tube. A typical cartridge fuse of this type is illustrated in Figure 2.14. The wire in the fuse is designed to carry a specific current before it begins to get hot. For the fuse in Figure 2.10 the current is 1 amp. If a current much higher than 1 amp is passed through the fuse, the thin

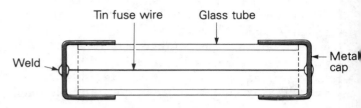

Figure 2.14 A typical cartridge fuse.

wire inside it gets so hot that it melts, breaking the circuit and interrupting the flow of electric current.

So if something goes wrong with the lamp-holder, causing the two terminals of the lamp to become connected together – 'shorted together' is the usual technical expression – the fuse will prevent damage to the wiring or to the battery by interrupting the current. Without the fuse, the wiring might melt or the battery might overheat, causing a fire.

Fuses are available in a range of values, and the circuit designer uses one that has a current-carrying capacity that is just a little more than the highest current that is likely to flow in the circuit when it is working properly.

The lamp itself can be a protection device of sorts. A lamp is used in the circuit in Figure 2.4 to limit the current. Normally, the amount of current that can flow is not enough to light the lamp, but in the event of a short-circuit (if the anode and cathode touch) the maximum current that can flow is limited by the lamp. The lamp also lights up, indicating that there is something wrong!

A disadvantage of a fuse is that, once it has 'blown', it is useless and has to be replaced. If a circuit is often subject to fault conditions, this is inconvenient and expensive. In such circuits an overcurrent **circuit-breaker** could be used. This is a device that interrupts the flow of a current by

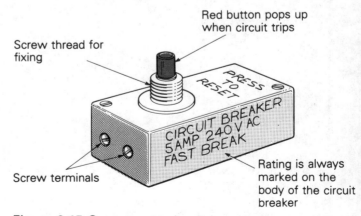

Figure 2.15 Overcurrent circuit-breaker.

opening a switch when the current increases above a certain level. Once the fault is corrected, the circuit-breaker can be reset by simply pressing a button. An overcurrent circuit-breaker is illustrated in Figure 2.15.

Having described the main components of a typical circuit, we can now begin to look at the way in which the parts interact with each other. The e.m.f., current and electrical resistance of the load are related in a simple mathematical way, and in the next section we shall look at what is meant by 'resistance', and at the relationship of these three factors.

Ohm's law

The flow of electric current through a circuit depends on two factors: the e.m.f. and the resistance of the circuit. To get a visual picture of resistance, it is convenient to think of the electric circuit as a plumbing system. Figure 2.16 shows just such a comparison.

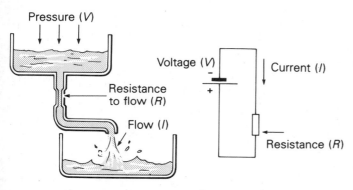

Figure 2.16 Plumbing analogy of an electric circuit: voltage, current and resistance all have their equivalents in the water system.

If the current flow is equivalent to a flow of water through the system, then the e.m.f. of the battery (in volts) is equivalent to the water pressure in the top tank (in kilograms per square metre). The flow of current (in amperes) in the circuit is equivalent to the flow of water in the pipe (in litres per minute). There is a restriction in the pipe that limits the flow. The amount of water that can flow out of the end of the pipe depends on the size of this restriction. If it is very thin, only a trickle of water will escape.

In the electrical system, the equivalent of the

restriction is a component called a **resistor** (because it resists the flow of electric current). The resistor has a greater resistance to the flow of current than the wires, just as the narrow part of the pipe 'resists' the flow of water more than the rest of the pipe. Without the resistor, a much larger current would flow in the circuit, just as more water would flow out of an unrestricted pipe. But notice that the amount of water would still not be unlimited; the pipe itself puts a limitation on the flow. It is the same in the electrical circuit, for the wires and even the battery exhibit a certain amount of resistance that would, in the absence of anything else, limit the current to some extent.

It is clear that, if the analogy holds good, there will be a relationship between pressure (e.m.f.), flow (current) and the size of the restriction (resistance). For example, if the water pressure were increased, you would expect a greater flow through the same restricted pipe.

The relationship between current, voltage and resistance was first discovered by George Simon Ohm in 1827. It is called **Ohm's law** after him.

Ohm's law states that the current (I) flowing through an element in a circuit is directly proportional to the p.d. (V) across it. Ohm's law is usually written in the form

$$V = IR$$

In words, this says that the voltage (in volts) equals the current (in amperes) times the resistance (in ohms). The unit of resistance, the **ohm** (Ω), is of course, named in honour of Ohm's discovery.

From the formula above we can see that a p.d. of 1 volt causes a current of 1 amp to flow through a circuit element having a resistance of 1 ohm. Given any two or three factors, we can find the other one. The formula can be rearranged as

$$I = \frac{V}{R}$$

or

$$R = \frac{V}{I}$$

This simple formula is probably used more than

any other calculation by practical electrical and electronics engineers. Given a voltage, it is possible to arrange for a specific current to flow through a circuit by including a suitable resistor in the circuit.

For some applications in electrical engineering and for most applications in electronic engineering, the ampere and the volt are rather large units. The ohm, by contrast, is rather a small unit of resistance. It is normal for these three units to be used in conjunction with the usual SI prefixes to make multiples and submultiples of the basic units. A chart of these is given in Table 2.1.

Table 2.1 SI prefixes.

Prefix	Symbol	Meaning	Pronunciation
tera	T	$\times 10^{12}$	tare-ah
giga	G	$\times 10^{9}$	guy-ger
mega	M	$\times 1\ 000\ 000$	megger
kilo	k	$\times 1\ 000$	keel-oh
milli	m	$\div 1\ 000$	milly
micro	μ	$\div 1\ 000\ 000$	–
nano	n	$\div 10^{9}$	nar-no
pico	p	$\div 10^{12}$	peeko
femto	f	$\div 10^{15}$	femm-toe

When working out Ohm's law calculations, it is vital that you remember to work in the right units. You cannot mix volts, ohms, and milliamps!

Kirchhoff's laws

In the 1850s, Gustav Robert Kirchhoff formulated two more laws relating to electric circuits. These laws, named after him, enable us to write down equations to represent the circuits mathematically.

Kirchhoff's laws:
• The sum of the currents flowing into any junction in a circuit is always equal to the sum of the currents flowing away from it.
• The sum of the potential difference in any closed loop of a circuit equals the sum of the electromotive forces in the loop.

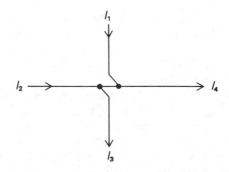

Figure 2.17 The junction between four current-carrying wires.

Let us begin by considering the first of Kirchhoff's laws. Figure 2.17 shows four wires, all carrying current and all connected together. This is the sort of situation that occurs in almost any piece of electrical equipment. In Figure 2.17, there are two wires through which current flows into the junction and two through which current flows away from it. Adding together the currents flowing in, we shall get exactly the same value as we shall for current flowing out. For Figure 2.17 this can be written as an equation:

$$I_1 + I_2 = I_3 + I_4$$

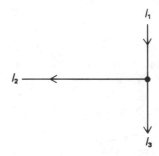

Figure 2.18 The junction between three current-carrying wires.

Figure 2.18 shows a three-wire junction, in which one wire carries current into the junction and two carry current away. The equation for this junction is

$$I_1 = I_2 + I_3$$

What Kirchhoff is saying is rather simple and obvious: current has come from somewhere, and it has to go somewhere; it can't just disappear.

Now for the second of Kirchhoff's laws. Figure 2.19 shows a simple circuit consisting of a source of e.m.f. and two resistors. In Figure 2.19, the source of e.m.f. is a 12 volt battery, and the resistors

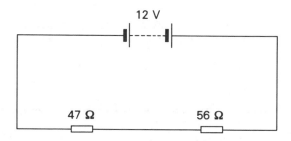

Figure 2.19 A simple circuit containing two resistors.

have values of 47 Ω and 56 Ω. We can use Ohm's law to determine the total current flowing in the circuit:

$$I = \frac{V}{R}$$

$$I = \frac{12}{47 + 56}$$

$$I = \frac{12}{103}$$

$$I = 0.1165 \text{ amperes}$$

We can use Ohm's law again to calculate the p.d. across each resistor:

For the first: $V = IR$
                   $V = 0.1165 \times 47$
                   $V = 5.476 \text{ volts}$

For the second: $V = IR$
                   $V = 0.1165 \times 56$
                   $V = 6.524 \text{ volts}$

The sum of the p.d.s across the resistors is

$$5.476 + 6.524 = 12 \text{ volts}$$

which is just what Kirchhoff's second law predicts.

If you want to verify the values experimentally, you could construct the circuit in Figure 2.19, using a 12 volt battery and two resistors of 47 Ω and 56 Ω, then measuring the e.m.f. and p.d.s with a suitable voltmeter. Is the result *exactly* what you were expecting?

Let's look at a more complicated example in a circuit with two sources of e.m.f. and three resistors (Figure 2.20). Start at the top left-hand corner of the circuit and work round, measuring all the voltages. The first is E_1, then E_2. Now do the same thing for the resistors R_1, R_2 and R_3. This provides you with the two sides of the equation, which, written out in full, is

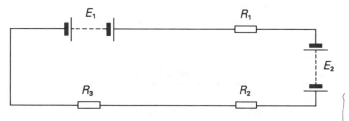

Figure 2.20 A circuit with three resistors and two sources of e.m.f.

$$E_1 + E_2 = R_1 + R_2 + R_3$$

You will of course get the same result by starting at any point in the circuit, provided you go right round the loop, back to the starting point.

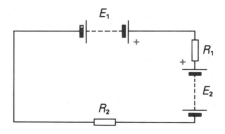

Figure 2.21 A circuit with opposing sources of e.m.f.

Figure 2.21 is similar, but note that one of the sources of e.m.f. is connected so that it opposes, rather than augments, the current flow; it could be a battery connected 'backwards' in the circuit. The way to deal with this is simply to give one of the sources of e.m.f. the opposite sign: to make it negative. It doesn't matter which is which, provided you make all the sources that face one way positive, and all the others negative. The equation is:

$$E_1 - E_2 = R_1 + R_2$$

Kirchhoff's laws are useful in allowing engineers to analyse circuits in detail.

Passive components: resistors

A resistor is a component that is used in electronic or electrical apparatus. It is designed to have a specific amount of resistance (measured, of course, in ohms). As we shall see, there are several kinds of resistor. Resistors can be obtained in a wide range of values. Before looking at the components themselves, we must consider the factors that have to be taken into account when specifying a resistor in a circuit.

Resistors in series and in parallel

It is possible to connect more than one load to a source of e.m.f., as we have seen above. Figure 2.22 shows two ways in which two loads, in the shape of resistors, can be connected to a battery. These resistors have resistances of 10 Ω and 5 Ω.

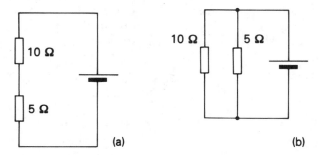

Figure 2.22 Resistors (a) in series and (b) in parallel.

In Figure 2.22(a), where current flowing through one load also flows through the other, the loads are said to be connected in **series**. In Figure 2.22(b), current flows through each load independently. The loads are said to be connected in **parallel**.

How can we calculate the combined value of resistance, as 'seen' by the battery? For resistive loads connected in series, the values are simply added together.

> The combined resistance, R_t, of the loads connected in series is given by the simple formula
>
> $$R_t = R_1 + R_2 + R_3 + \ldots$$
>
> (the dots mean that you can add as many numbers as you like).

In Figure 2.22(a), this is

$$R_t = 10 + 5$$
$$R_t = 15$$

which is about as simple as you can get. The second case, illustrated in Figure 2.22(b), is less easy to calculate.

> The formula for working out a combined parallel resistance is
>
> $$\frac{1}{R_t} = \frac{1}{R_1} + \frac{1}{R_2} + \frac{1}{R_3} \ldots$$

This means that the reciprocals of the values of the resistances, added together, give the reciprocal of the answer. Working this out for Figure 2.22(b) we get

$$\frac{1}{R_t} = \frac{1}{10} + \frac{1}{5}$$

$$\frac{1}{R_t} = 0.1 + 0.2$$

$$\frac{1}{R_t} = 0.3$$

$$R_t = 3.33$$

The combined resistance of the 10 Ω resistor and 5 Ω resistor, connected in parallel, is 3.3 Ω. Reciprocals can be obtained with reciprocal tables or, more usually, with a pocket calculator.

Accuracy of calculations

My pocket calculator shows that the reciprocal of 0.3 is 3.333333333. This seems to be a very high degree of accuracy, but the accuracy of the answer (not the accuracy of the reciprocal) is unreal when applied to the resistor. Electronics is mostly about building or repairing circuits in the real world, where nothing (not even an electronic component!) is perfect.

We must consider how accurately we know the true values of the two resistive loads we started with. Is the first one really 10 Ω to ten decimal places? This is most unlikely. It is neither necessary nor possible for manufacturers to make a resistor so carefully. We should also consider how accurately we *need* to know the answer. For most purposes, and bearing in mind the very large number of variables that exist in any electronic circuit, it is usually quite enough to say that the combined parallel resistance is about 3.3 Ω.

This doesn't mean that you can be slipshod about calculations. Work out your calculations accurately and *then* decide (or, in some cases, calculate) how accurate the answer is likely to be. Or how accurate you need it to be. Electronic circuits seldom include any absolute values; nothing is *exactly* what it claims to be.

In practice, you may come across quite complex combinations of series and parallel loads. Figure 2.23 illustrates just such a combination. The rule for dealing with part of a circuit like this, and arriving at the combined resistance of all the loads, is to work out any obvious parallel and series combinations first, progressively simplifying the circuit. There is an obvious parallel combination in this figure, so we can work out the combined resistance of the two 100 Ω and two 50 Ω resistances first. Using a calculation like the one above, we arrive at a total resistance of about 16.7 Ω for the four parallel resistances. Figure 2.23 has now become more simple: the simplified version is shown in Figure 2.24(a).

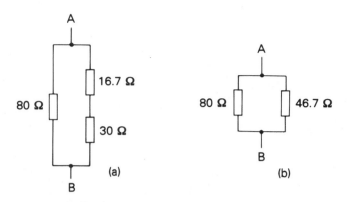

Figure 2.24 Progressive simplifications of Figure 2.23.

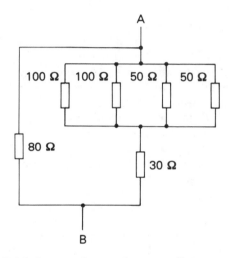

Figure 2.23 A complex series–parallel network.

The two series resistances, 16.7 Ω and 30 Ω, can be added together, to give the simpler circuit in Figure 2.24(b). This leaves us with a final parallel calculation to make, which gives an answer of 29.5 Ω or, in the real world, about 30 Ω.

Work through this example, then change some of the values and try again. Make sure that you are completely confident that you can perform this kind of calculation without error. All calculations like this are a lot simpler if you use a pocket calculator, preferably one that will work out reciprocals. Calculators are inexpensive and any student of electronics should have one handy.

Tolerances and preferred values

It should be clear by now that a load having a given resistance of, say, 20 Ω is unlikely to have a resistance of *exactly* 20 Ω if you measure it accurately enough. Components having a specified resistance, such as resistors used in electronic circuits, may be marked with a specific value of resistance. But inevitably the components will vary slightly, and not all resistors stamped '20 Ω' will have a resistance that is exactly 20 Ω. Manufacturing tolerances may be quite wide.

When designing or repairing circuits it is important to bear in mind that, because of this, components may vary quite a lot from their marked values. As we saw above, this affects the calculations we make, and in any complex circuit we would have to specify circuit values of resistance, voltage and current in terms of a range of values. A resistor marked '20 Ω' might, for example, have a resistance of 18 Ω or 22 Ω. These two figures represent a departure from the marked value of around ± 10 per cent. This is in fact a typical manufacturing tolerance for resistors.

Let us look at the effect of two such resistors connected in a series circuit (Figure 2.25). Using simple addition they add up to 50 Ω. We might base our circuit design on this and might, for example, predict that this very simple circuit would draw 100 mA if connected to a 5 V supply (see Ohm's law above).

However, let us consider the two extremes. If both resistors happened to be at the upper limit of their tolerance range, the combined resistance could be 55 Ω. If they were both at the lower limits, then a combined value of 45 Ω might be expected. The circuit could therefore be taking a

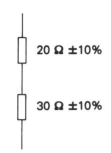

20 Ω ±10%

30 Ω ±10%

Figure 2.25 The effect of tolerance in values.

current of 90 mA (lowest) or 108 mA (highest). According to the function of the circuit this might or might not be important. Statistically, the more components there are in a circuit, the more likely that circuit is to conform to the 'average', or marked, values overall.

Table 2.2 Preferred values of resistors.

E12 series	E24 series
10	10
	11
12	12
	13
15	15
	16
18	18
	20
22	22
	24
27	27
	30
33	33
	36
39	39
	43
47	47
	51
56	56
	62
68	68
	75
82	82
	91

E12 series is available in all types of resistor.
E24 series is available in close-tolerance or high-stability only.

All values are obtainable in multiples or submultiples of 10, e.g. 2.2 Ω, 22 Ω, 220 Ω, 2.2 kΩ, 22 kΩ, 220 kΩ, 2.2 MΩ, 22 MΩ.

Resistance values less than 10 Ω and more than 10 MΩ are uncommon, and may not be available in all types of resistor.

In some cases we need to be able to specify a resistor more closely than ± 10 per cent. Manufacturers of these components therefore produce different ranges of resistors, according to the degree of tolerance allowed. The closer the tolerance, the more expensive the resistor, so it is commercially unwise to specify tolerances closer than are necessary for correct operation of any given circuit. It is also impossible for manufacturers to produce all possible values of resistor (imagine how big their catalogues would have to be!), so they restrict themselves to an internationally agreed range of standard values, known as **preferred values**. Table 2.2 lists those available.

There are two ranges available, known as the E12 series and the E24 series. E24 resistors are usually only available in tolerance ranges less than ± 10 per cent. Intermediate values are not generally available, but can be made by combining the preferred values. It is important to think before doing this: there is little purpose in combining a 47 Ω resistor with a 3.3 Ω resistor in the hope of ending up with one that is exactly 50 Ω, if the tolerance of the combinations is likely to be ± 5 Ω.

There are usually three tolerance ranges available from most manufacturers: ± 5 per cent (which is the most common these days), ± 10 per cent, and 'close tolerance', which might be ± 2.5 per cent or even ± 1 per cent. Components to this standard of accuracy would normally be used only in measuring equipment or in particularly critical parts of some circuits.

Types of resistor

The most usual type of resistor is the **solid carbon resistor** (see Figure 2.26). Its structure is very simple; it consists of a small cylinder of carbon,

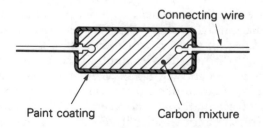

Connecting wire

Paint coating   Carbon mixture

Figure 2.26 A solid carbon resistor (this component might be anything from a few millimetres to a few tens of millimetres long).

which is mixed with a non-conductor. A connecting wire is fixed into each end, and the resistor is given a coat of paint to protect it from moisture, which might alter the resistance.

Resistors are always marked with a colour code to indicate the value. The colour code consists of three or four coloured bands painted round the resistor body. This system is used because it makes the resistor's value visible from any direction; a printed label could be hard to read with the component in place on a crowded board. Also, a painted or printed value could easily get rubbed off, whereas painted bands are relatively permanent.

The first three bands of the colour code represent the value of the resistor in ohms. Bands 1 and 2 are the two digits of the value, and band 3 represents the number of zeros following the first two digits. The fourth band is used to indicate the tolerance of the resistor's stated value. Figure 2.27 gives the resistor colour code; the standard is international.

Where values of resistance less than two digits are required, the two bands are followed by a gold band. Thus 4.7 Ω would be represented as yellow, purple, gold plus a tolerance band.

The next kind of resistor is the **metal oxide** or **metal glaze resistor**. This looks rather like the carbon resistor from the outside, but the internal

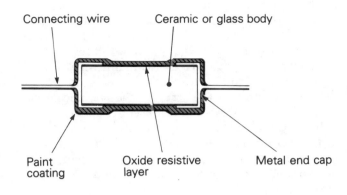

(a)

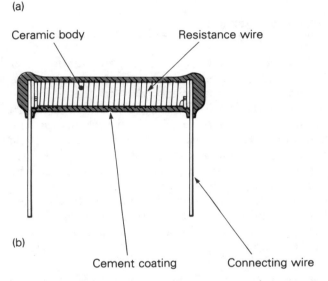

(b)

Figure 2.28 (a) A metal oxide resistor. (b) For high-power applications, a wire-wound resistor is used.

structure is different. Figure 2.28(a) shows a cross-section. Metal oxide resistors can be made to closer tolerances than carbon resistors, and change their resistance less with changes in temperature. For this reason they are sometimes called **high-stability resistors**. The resistance of a metal oxide resistor changes approximately 250 parts per million per °C. This compares with about 1200 parts per million per °C for carbon resistors.

Both carbon and metal oxide resistors are made in a range of stock sizes, from about 0.125 watts dissipation to about 3 watts (see the section on power dissipation later in this chapter for the definition of watts). Sometimes it is necessary to have resistors that can cope with higher powers; for this, wire-wound resistors are used. The **wire-wound resistor** is shown in Figure 2.28(b).

It is easy to make wire-wound resistors in low resistance values, down to fractions of an ohm.

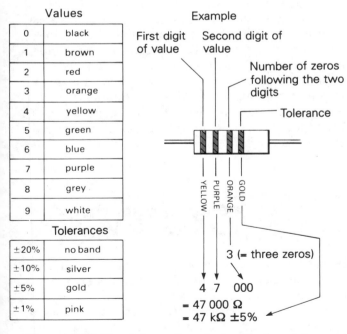

Figure 2.27 The international standard resistor colour code.

High resistance values use wire of low conductivity, requiring many turns of fine-gauge wire as well. The maximum practical value for a wire-wound resistor is a few tens of kilohms, at least for components that are a reasonable size. Power ratings range from 1 to 50 W in the stock sizes; there is no limit to the size in practice, and larger special-purpose wire-wound resistors are common.

It is possible to make precision wire-wound resistors in which the resistance is specified to a very close tolerance, within ±0.1 per cent. Such resistors are expensive and would be used only in measuring equipment.

Variable resistors

For applications such as volume controls and other 'user controls' in electronic equipment it is often necessary to have a resistor that can be altered in resistance by means of a control knob. Such resistors are called variable resistors, or **potentiometers**. They consist of a resistive track, made with a connection at either end. A movable brush, generally made of a non-corroding metal, can be moved along the track; an electrical connection to the brush allows a variable resistance to be obtained between either end of the track and the brush. Figure 2.29 shows the main components of a typical potentiometer.

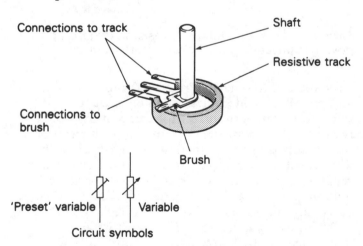

Figure 2.29 Principle of the potentiometer or variable resistor.

Three forms of variable resistor are shown in Figure 2.30. Note that the shaft (or slot) that moves the brush is generally connected to the brush, without any insulation; it is important to know this for safety reasons.

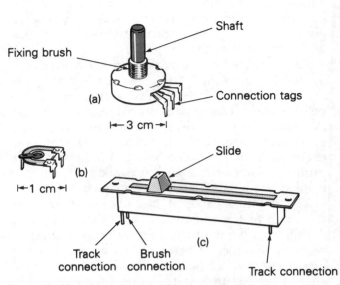

Figure 2.30 Three different types of potentiometer.

Variable resistors are available in various shapes and sizes, with power dissipations from around 0.25 W upwards. Tracks are either carbon, conductive ceramic ('cermet') or wire-wound. Resistance ranges are available between fractions of an ohm and a few megohms.

The track can be made in such a way that the resistance increases smoothly along the track: the usual linear type of potentiometer. For some audio uses (such as volume controls) logarithmic potentiometers are made in which the resistance increases according to a logarithmic law rather than a linear law. For most purposes you need only remember that the way logarithmic potentiometers control volume approximates to the way the human ear responds to sounds of different loudness. If you use a linear potentiometer for a volume control, the effect of the control seems to be 'all at one end' of the scale.

Passive components: capacitors

Next to resistors, the most commonly encountered components are capacitors. A capacitor is a component that can store electric charge. In essence, it consists of two flat parallel plates, very close to each other, but separated by an insulator (see Figure 2.31). When the capacitor is connected to a voltage supply, a current will flow through the circuit (see Figure 2.32). Electrons are stored in one of the plates of the capacitor; in the other there is a shortage of electrons. In this state the capacitor is said to be **charged**, and if it is disconnected from the supply, the imbalance in potential between the two plates will remain.

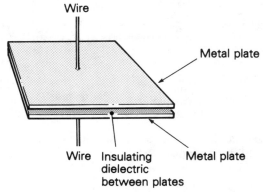

Figure 2.31 Schematic view of a capacitor.

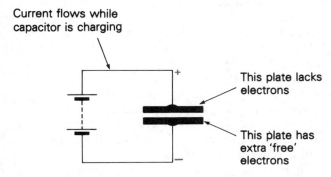

Figure 2.32 A capacitor connected to a power supply; current will flow until the capacitor is charged.

If the charged capacitor is connected in a circuit it will, for a short time, act as a voltage source, just like a battery. This can be demonstrated quite nicely with nothing more than a large capacitor rated at 15 V, 1000 μF (see below regarding units of capacitance). Note that the capacitor will be an electrolytic type (again, see below) and must be connected to the battery with its + terminal towards the positive (+) terminal of the battery. The demonstration is shown in Figure 2.33.

Try the experiment again, this time allowing 2 minutes between charging the capacitor and discharging it into the bulb. Try waiting progressively longer, and you will see that the charge gradually leaks away on its own. This is due to the imperfection of the insulator separating the plates, allowing a tiny leakage current to flow. A graph of the

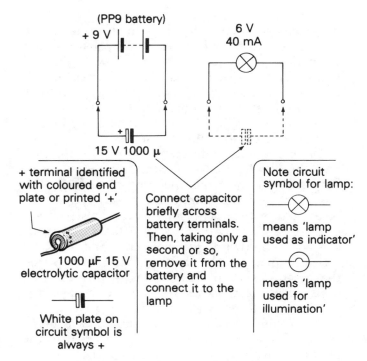

Figure 2.33 *A simple experiment for demonstrating the way in which capacitors can store power. A large capacitor is first charged up from a 9 volt supply, and then discharged into a small lamp.

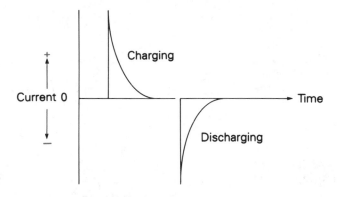

Figure 2.34 A graph showing a capacitor charging and discharging.

current flowing through a charging and then discharging capacitor is compared with the voltage measured across it in Figure 2.34.

The unit of capacitance is the **farad**, named after Faraday. The farad is a huge unit of capacitance in the context of electronic circuits, and the smaller derived units of capacitance are used. The largest practical unit is the microfarad, symbol μF. Although values of capacitance in the range of thousands of microfarad are sometimes used, the millifarad is never encountered: a capacitor would be marked 10 000 µF, for example, not 10 mF. If you do see a capacitor marked 'mF', it is almost certainly meant to be µF, with that manufacturer using a non-standard abbreviation. This is becoming less common. In addition to the µF, the nF and the pF are commonly used units.

There are many different types of capacitor, according to the use to which the component is to be put, and also to the operating conditions. The capacitance of the component is determined by three factors: the area of the plates, the separation of the plates, and the insulating material that separates them, known as the **dielectric**. For the same dielectric material, the closer and larger the plates, the greater the capacitance. Another factor that determines the thickness of the dielectric is the maximum voltage to which it is going to be subjected. If the dielectric is very thin, it will break down with a relatively low voltage applied to the plates of the capacitor. Once the dielectric has been damaged, the capacitor is useless. High-voltage capacitors need thicker dielectrics, to withstand the higher voltage. To produce the same capacitance, the plates have to be larger in area, so the component is bigger.

The simplest type of capacitor uses a roll of very thin aluminium foil, interleaved with a very thin plastic dielectric such as Mylar. A physically smaller capacitor can be made by actually plating the

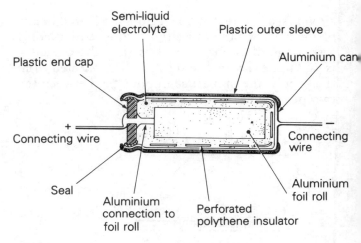

Figure 2.36 Construction of an aluminium electrolytic capacitor.

aluminium onto one side of the Mylar (see Figure 2.35). For higher voltages, polyester, polystyrene or polycarbonate plastic material is used as the dielectric.

Ceramic capacitors are used where small values of capacitance and large values of leakage current are acceptable; the ceramic capacitor is inexpensive. A thin ceramic dielectric is metallised on each side, and coated with a thick protective layer, usually applied by dipping the component.

Both plastic film and ceramic capacitors are available in the range 10 pF to 1 µF, although plastic film types may be obtained in larger values.

Where really high values of capacitance are needed, **electrolytic capacitors** are used. Electrolytic capacitors give very large values of capacitance in a small component, at the expense of a wide tolerance in the marked value (−25 to +50 per cent) and the necessity for connecting the capacitor so that one terminal is always positive.

The most commonly used type of electrolytic capacitor is the aluminium electrolytic capacitor. The construction is shown in Figure 2.36. After manufacture, the capacitor is connected to a controlled current source, which electrochemically deposits a layer of aluminium oxide on the surface of the 'positive' plate. The aluminium oxide makes an excellent dielectric, with very good dielectric strength (resistance to voltage applied across the plates). Since the layer is chemically deposited it is very thin.

The electrolytic capacitor must not be subjected to voltages applied in the wrong direction, or the aluminium oxide layer will be moved off the positive plate, and back to the electrolyte again.

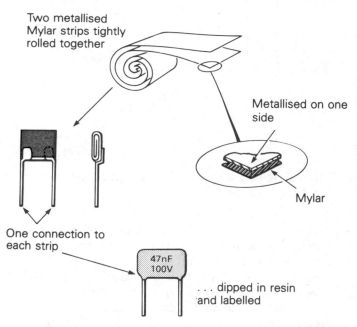

Figure 2.35 A typical capacitor, used in electronics work: the 'metallised film' type.

Variable capacitors

Capacitors are made that can be varied in value. Either air or thin mica sheets are used as the dielectric, so the capacitance of variable capacitors is usually low. Rotary or compression types are made. In the rotary type, the capacitance is altered by changing the amount of overlap of the two sets of plates (Figure 2.37). In compression trimmers, the spacing between plates is altered, as in Figure 2.38.

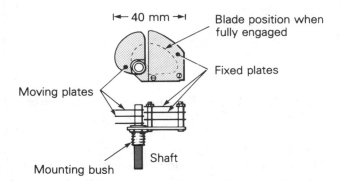

Figure 2.37 A variable capacitor of the 'air space' type. This component would be used for tuning a radio receiver, for example.

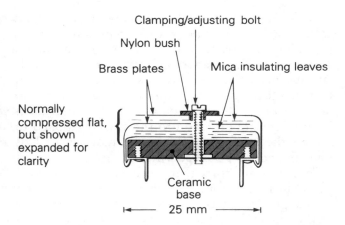

Figure 2.38 A 'compression trimmer': a form of variable capacitor used to preset circuit values (adjustment is by means of a screw).

Variable capacitors can be obtained with maximum values from 2 pF to 500 pF.

Capacitors as frequency-sensitive resistors

Once a capacitor has charged up, it will not pass current when connected to a direct voltage supply.

However, if we reverse the polarity of the supply, the capacitor will permit a current to flow until it has charged up again, with the opposite plates positive and negative. The capacitor will then, once again, block the flow of current. If the frequency of an alternating voltage applied to a capacitor is high enough the capacitor will actually behave as if it were a low-value resistor, as it will never become charged in either direction. At lower frequencies, the capacitor will appear to have a higher resistance, and at zero frequency (d.c.) it will, as we have seen, have an infinitely high resistance if you disregard the leakage current.

The precise value of resistance shown by the capacitor will depend upon the capacitance, the applied voltage and the frequency. The property is known as **reactance**. It enables the circuit designer to use the capacitor to block or accept different frequencies in a variety of different circuit configurations, many of which you will meet later in this book.

Capacitors in series and parallel

Capacitors can be used in series and in parallel. When used in series, the working voltage is the sum of the two working voltages, so two capacitors intended for a 10 V maximum supply voltage could be used, in series, with a 20 V supply.

The calculation of capacitor values in series and parallel is similar to the calculation used for resistors in series and parallel, but the opposite way round. Thus for capacitors in parallel the formula

$$C_t = C_1 + C_2 + C_3 + \ldots$$

is used, simply adding the values together. For capacitors in series,

$$\frac{1}{C_t} = \frac{1}{C_1} + \frac{1}{C_2} + \frac{1}{C_3} \ldots$$

gives the total value of capacitance.

Unfortunately, there is no internationally agreed colour code for capacitors, so manufacturers usually stamp the value on a capacitor, using figures in the normal way. A few capacitors are marked with coloured bands, but it is wise to look at the maker's catalogue to check what the code means.

Passive components: inductors

The third main passive component is the inductor. This is a coil of insulated wire that may, or may not, be wound over a ferrous metal former. When a coil is connected in a circuit, as in Figure 2.39, the flow of current through the coil causes an electromagnetic field to be created around the coil. Building the field absorbs energy. This changing (increasing in strength) magnetic field produces its own current in the coil, opposing the direction of the applied current.

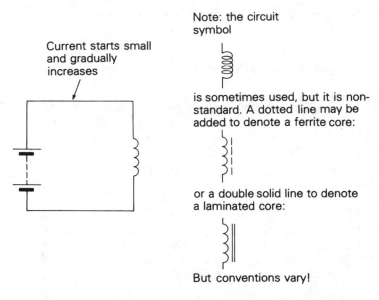

Figure 2.39 An inductor connected in a simple circuit; the current rises to a fixed level, determined by the resistance of the inductor's windings.

The effect is that, when current is first applied to the inductor, the inductor seems to have a high resistance, reducing the current flow through it. Once the electromagnetic field is established, the resistance drops. A graph of the current flowing through an inductor, and the voltage measured across it, is given in Figure 2.40.

Compare the graph with the one in Figure 2.34, and you will see that the electrical characteristics of an inductor are the 'opposite' to those of a capacitor! Once the current flowing through the inductor is constant, the only opposition to current

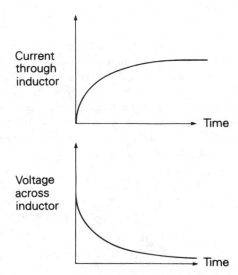

Figure 2.40 Graphs showing voltage and current in the circuit of Figure 2.39.

flow comes from the resistance of the coil wire. This is usually small compared with the apparent resistance (called **inductive reactance**) when the current is building up. All the while that a steady current is flowing in the inductor, the magnetic effect of a current means that there will be a steady magnetic field associated with the inductor.

If the current is switched off, the magnetic field collapses; but the energy that went into it can't just disappear (energy never does), it has to go somewhere. What happens is quite simple. The collapsing magnetic field causes a p.d. to appear between the ends of the coil; if a circuit were connected to it, a current would flow. This current is in the opposite direction to the current applied to set up the magnetic field in the first place. This phenomenon is called **self-inductance**.

The opposing force that prevents current flowing through an inductor right away is **induced current**. A changing magnetic field will induce a current in **any** conductor in the vicinity; when a current is induced in a conductor **other** than the one carrying the current that was responsible for the magnetic field, the effect is called **mutual inductance**.

The unit of inductance is the **henry**, symbol H. One henry is defined as the inductance that will produce an induced e.m.f. of 1 volt when the current through it is changing at the rate of 1 ampere per second. In electronics, mH and μH are commonly encountered but the henry, like the farad, is rather a large unit.

Inductors are available as a range of components in small inductance values from a few μH to a few mH. However, there are so many variable factors that a 'stock range' of all types would be too large for a supplier to produce. Where inductors are needed (and they are relatively rare in modern circuits) they may well be specified in terms of such things as wire thickness, number of turns and core type.

Inductors can be used in series or parallel combinations, and the combined value of inductance is calculated just as you would for resistors in series or parallel (given earlier in this chapter).

Transformers

Transformers make use of mutual inductance, in which a current flowing in a coil produces an electromagnetic field, which in turn induces a current to flow in a second coil wound over the first one.

The construction of a transformer is shown in Figure 2.41. The iron laminated core is used to concentrate the electromagnetism and thus improve the efficiency. It is important that the iron laminations are insulated from each other; if they are not, the core itself will behave as if it were a one-turn coil, and a current will be induced in it. The current would be very large, and cause the transformer to overheat.

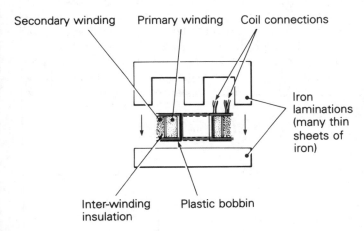

Figure 2.41 A typical small transformer, showing the iron laminations.

The ratio of the input voltage to the primary winding to the output voltage from the secondary winding depends on the **turns ratio** of the coils, and is approximately equal to that ratio. Thus a transformer with 2000 turns on the primary and 1000 turns on the secondary will have an output voltage that is half the input voltage.

Since currents are induced by **changing** magnetic fields, the transformer will operate only if the input voltage is constantly changing. The transformer could be used with alternating current (such as the mains supply), and one of the common uses for a transformer in electronics is to reduce the voltage of the a.c. mains to a lower level, suitable for electronic circuits. This is illustrated in Figure 2.42.

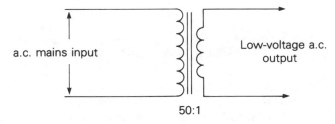

Figure 2.42 Circuit symbol for a transformer with a laminated core.

Transformers are rated according to the turns ratio, the power-handling capability (in watts), and the type of application. Power transformers, intended for power supplies, usually have mains voltage primary windings and a high standard of insulation between the primary and secondary coils, for safety. The secondary winding may have a range of connections, called **taps**, so that the transformer can be used in different applications.

> It is important to realise that the transformer simply transforms voltages; there is no net power gain, always a small loss.

If a transformer has a 200 V primary and a 1000 V secondary, the voltage will be increased by a factor of five times, but it will require slightly more than five times the current in the primary than will be available from the secondary. You can trade current for voltage and vice versa, but the power will always be slightly less at the output, owing to various losses in the transformer (dissipated, as usual, in the form of heat).

Any component having a coil that carries current will have an inductive characteristic, though it may be swamped by other factors (resistance or capacitance). This may sometimes be important. For example, a **relay** (see Chapter 7) is an electromagnetic switch, the coil being used to operate a mechanical switch. However, the coil has inductance. A relay operating from a 6 V supply can, when suddenly disconnected, produce an output pulse of 100 V or more as the current flowing through it drops very rapidly to zero. This voltage may be high enough to harm semiconductor components, and precautions have to be taken to ensure that this 'spike' of high voltage is made harmless. More about this later.

Power dissipation

We have seen that resistors insert a certain amount of resistance into a circuit, restricting the current flow. The simple Ohm's law calculation below shows one example that could apply to a simple circuit.

Supply voltage = 24 V
Current = 240 mA
Resistor = 100 Ω

$$V = IR$$
$$24 = \frac{240}{1000} \times 100$$

When discussing the various factors involved in specifying a resistor, you may remember we mentioned power dissipation. This is an important factor in any circuit, as you will see. If you were to connect a 100 Ω resistor across a 24 V power supply, you would quite quickly notice something about the resistor: it gets hot. In this particular circuit, the resistor would quite quickly get too hot to touch. Why?

The answer is that power from the supply is dissipated (lost) to the surroundings by the resistor. The power is lost in the form of heat, as you will remember from reading the first part of this chapter (Electrical effects). It is easy to calculate just how much heat is lost.

If we were to measure the p.d. across the resistor, we would find that it is 24 V. Since the two ends of the resistor are connected to the two power supply terminals, this is only to be expected. The current flowing is 240 mA. It can be calculated using Ohm's law, or measured directly. The amount of energy dissipated by this circuit can be calculated by the simple formula

$$P = VI$$

where P represents the power in **watts**, symbol W.

The watt is a unit of energy, and (as illustrated above) is an amount of energy equal to that dissipated when a current of 1 ampere is flowing across a potential difference of 1 volt.

Thus, in our circuit, the amount of power dissipated is

$$24 \times \frac{240}{1000} = 5.76 \text{ watts}$$

which is enough to be quite hot to touch.

A large resistor can dissipate more heat than a small one. The 'usual' size of resistor used in small items of electronic equipment such as radios and tape recorders is 0.25 W. A resistor of this size, if connected in the circuit we are considering, would quickly overheat and burn through at 5.76 watts. A larger, wire-wound component would have to be used, and even then it would be necessary to take precautions to allow the heat to escape: perhaps the casing of the equipment would have ventilation holes.

In electronic circuits, resistors with a power dissipation in excess of 1 watt are uncommon; designers try to use the minimum amount of power, and to waste as little as possible, particularly if the equipment is battery-powered.

The same formula as that given above is used to calculate the power dissipation of series and parallel combinations of resistors. Figure 2.43 shows such circuits. Figure 2.43(a) shows two resistors connected in series. It is quite clear that the same current flows through both resistors. However, the p.d. measured across each resistor will be different. The total p.d. is given as 5 volts. We can find the voltage across each individual resistor by using Ohm's law.

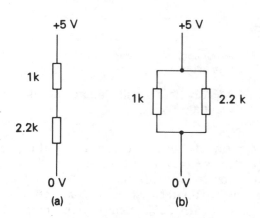

Figure 2.43 Resistors (a) in series and (b) in parallel.

We know that the total resistance of the circuit is 3.2 kΩ and can use Ohm's law to calculate that the current flowing through the circuit is about 1.562 mA.

Using the form of Ohm's law $V = IR$ we can calculate the voltage drop across each resistor. For the first resistor:

$$V = \frac{1.562}{1000} \times 1000$$

$$V = 1.56$$

And for the other resistor:

$$V = \frac{1.562}{1000} \times 2200$$

$$V = 3.44$$

Now we know the p.d. across each resistor, we can calculate the amount of power that each resistor is dissipating. For the 1 kΩ resistor this is

$$P = 1.56 \times 1.56$$
$$P = 2.43 \text{ mW}$$

And for the 2.2 kΩ resistor:

$$P = 1.56 \times 3.44$$
$$P = 5.37 \text{ mW}$$

Look carefully at the units in these calculations. The amounts of power dissipated are very small: typical, in fact, of the sort of thing that you will find in most electronic circuits. The smallest commonly available resistor (0.125 W) is a lot larger than is necessary to dissipate this amount of power.

Exactly the same sort of calculation can be used to work out the power radiated by resistors in a parallel circuit. Figure 2.43(b) shows such a circuit. In this case you can tell (just by looking at the circuit) that the p.d. across the two resistors will be the same, but that the current flowing through them will be different. Using Ohm's law to arrive at the current flowing through each resistor, we get

$$\text{(i)} \quad I = \frac{5 \times 1000}{1000}$$
$$I = 5 \text{ mA}$$

$$\text{(ii)} \quad I = \frac{5 \times 1000}{2200}$$
$$I = 2.27 \text{ mA}$$

We can now use the power calculation to determine the power dissipation of each:

$$\text{(i)} \quad P = 5 \times 5$$
$$P = 25 \text{ mW}$$
$$\text{(ii)} \quad P = 5 \times 2.27$$
$$P = 11.35 \text{ mW}$$

Once again, think about units. Why is the above answer in milliwatts, and not watts?

Circuit diagrams

We have already used circuit diagrams in this book, and as a student of electronics you will quickly find that, almost without thinking, you are using the 'correct' diagrams. Different countries use slightly different symbols, but in general the 'language' of circuit diagrams is international, and a well-drawn circuit diagram will be understood by an engineer from anywhere in the world: quite an achievement in itself!

There are a few 'ground rules' for good circuit diagrams. Generally, it is conventional to draw the

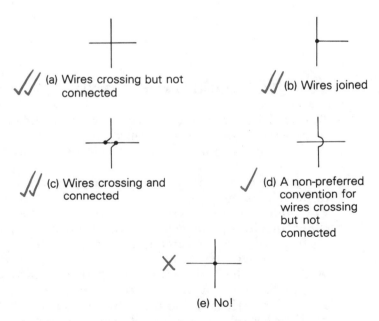

Figure 2.44 Circuit diagrams: (a) wires crossing but not connected; (b) wires joined; (c) wires crossing and connected; (d) a non-preferred convention for wires crossing but not connected; (e) something that should not appear on any circuit diagram.

diagram with the 'chassis' (or 'earthy') side of the power supply at the bottom. The 'live' supply rail will be at the top of the diagram, which makes the drawing easy to read for an engineer unfamiliar with the circuit. Most connecting wires are either horizontal or vertical, with only a few exceptions. Standard symbols are used for all components, and a good circuit diagram will have a clean, uncluttered appearance.

In encouraging students to get used to circuit diagrams, I shall introduce each new symbol as it comes up, rather than give a list at the beginning or end of the book. This will help to show how a symbol is used in context. For example, some circuit elements (such as multivibrator circuits, see Chapter 15) are almost always shown in the same, easily recognisable, form. This is something that cannot be learned from a list of symbols.

The only firm piece of graphical advice that I shall give at this stage is in the crossing and connecting of conductors in a circuit. Although international standards lay down quite clearly what is recommended, many engineers ignore this and produce drawings that are ambiguous. Figure 2.44 shows the way it should be done.

The circuit diagrams in this book all follow good standard practice, and can be used by students as a 'model' for the way diagrams ought to look.

■ CHECK YOUR UNDERSTANDING

● Electricity has various effects: heating, lighting, chemical, magnetic and mechanical.
● The causes and effects of electricity are reversible.
● Electric current is the movement of electrons through a conductor.
● **Conductors** are materials – usually metals – that allow an electric current to flow through them easily. **Insulators** are materials through which electric current will not flow.
● PVC sheathed wires are coded **red** or **brown** for live; **black** or **blue** for neutral; and **green, green and yellow stripes,** or **no insulation** for earth.
● In order to perform useful work, an electric current has to flow in a **circuit**, from the source of **electromotive force** (e.m.f.), through a **load**, then back again to the source.
● **Switches** are used to control the flow of current in a circuit; **fuses** or **circuit-breakers** are used to

prevent excessive current flowing if something goes wrong.
● **Ohm's law** relates voltage V (in volts), current I (in amps), and resistance R (in ohms): $V = IR$.
● **Kirchhoff's laws** enable us to represent circuits mathematically.
● **Resistors** are components designed to introduce a known amount of resistance into a circuit. Resistance is measured in ohms (symbol Ω). Resistors are made in many different values of resistance.
● **Capacitors** are components that store electric charge. Their capacitance is measured in farads (symbol F), or more often microfarads.
● **Inductors** introduce inductance into a circuit. It is measured in henrys (symbol H).
● **Transformers** increase or decrease alternating voltages.
● **Power** used in a circuit is measured in watts (symbol W). Power (P), voltage (V), and current (I) are related in the formula $P = VI$.

REVISION EXERCISES AND QUESTIONS

1 An electric current has a number of effects that can be demonstrated experimentally. What are they? Which ones can be reversed?
2 Describe how a piezoelectric record-player pick-up produces an electrical signal when it is playing a record.
3 Write down in full the following abbreviations:
   i) e.m.f.;
   ii) p.d.;
   iii) PVC;
   iv) μF;
   v) LED.
4 Mains-powered electrical equipment is always protected by a *fuse*. How does a fuse provide protection?
5 Write down the forms of Ohm's law that you would use:
   i) to calculate voltage, given current and resistance;
   ii) to calculate resistance, given voltage and current.
   iii) Write down the basic units that you would use for current, voltage and resistance when using Ohm's law.

6 Calculate the resistance across the terminals $T_1 - T_2$ of this network of resistors:

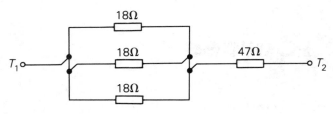

7 Write down Kirchhoff's first two laws.

8 Why are electrolytic capacitors used in some circuits instead of other kinds (such as metallised film capacitors)?

9 An electric kettle uses 1.5 kW when connected to a 240 V mains supply. What current flows through the kettle element? What rating of fuse would you use in a fused plug connected to the kettle lead?

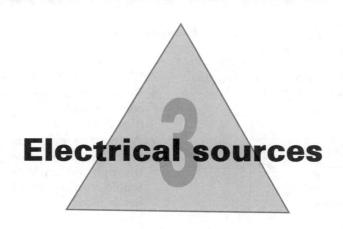

Electrical sources

Introduction

Almost everyone knows that batteries are sources of electric current. In most places electricity is supplied to houses and factories, and everyone understands that electricity can be produced in a power station by generators. In this chapter we shall be looking at all the common sources of electric current – there are more than you would think! – comparing them, and considering which are best for various applications.

Types of electric current

How can there be different types of electric current? This is a fair question. An electric current is always the same thing: a movement of electrons along a conductor. However, the movement need not be steady, and it need not be continuous. Nor need it always be in the same direction.

We have already considered a simple electric current, flowing in one direction only at a constant rate, when we looked at Ohm's and Kirchhoff's laws. Current that flows in a constant direction is called **direct current**, abbreviation d.c. Look at the

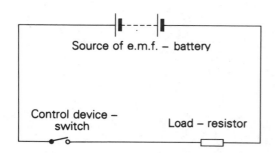

Figure 3.1 A simple switched circuit.

simple circuit in Figure 3.1. This shows a source of e.m.f. (a battery), a load (a resistor) and a control device (a switch). If the p.d. across the resistor is measured with the switch open, it will be zero. If it is measured with the switch closed, it will equal the battery voltage. If you were to open and close the switch at regular intervals, it would be possible to plot a graph of the way the p.d. across the resistor changed with time. The graph would look like Figure 3.2. The line of the graph is a visual representation of the p.d. across the resistor. The same line could also represent current flowing through the circuit. In electrical engineering, such a graph is usually called an **electrical waveform**.

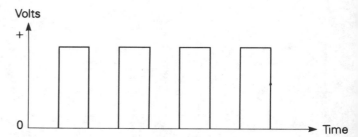

Figure 3.2 A graph of p.d. across the resistor in Figure 3.1 if the switch is alternately open and closed.

The current flowing in the circuit of Figure 3.1 is not continuous, but when it does flow, it always flows the same way. Such a current is termed **chopped d.c.**, or sometimes **pulsed d.c.** The circuit in Figure 3.3(a) is more complicated. It still shows a battery and a resistor, but this time a more complicated switch is used. The switch reverses the connections to the battery when it is operated, as illustrated in Figure 3.3(b).

Now imagine the waveform across the resistor, and how it would look when the switch is operated

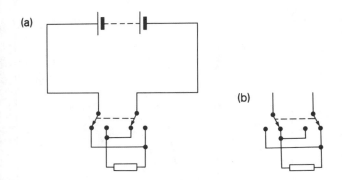

(a)

(b)

Figure 3.3 (a) A circuit with a double-pole switch. (b) Compare the double-pole switch in this position with 3.3 (a); follow the current path in both diagrams to see how the p.d. across the resistor is reversed by the switch.

continuously. The p.d. across the resistor would appear first one way round, and then the other. The polarity of the voltage would be continuously changing. The same would apply to the flow of current. Forgetting the mechanical details of the switch, and assuming that the switch operates instantaneously, the current is never zero, but it reverses continuously. Figure 3.4 shows the graph of the waveform. It might be current or voltage; it doesn't matter which. The important thing to notice is the fact that the centre line of the graph goes through the middle of the waveform, indicating that the p.d. across the resistor (or the current flow) continuously reverses.

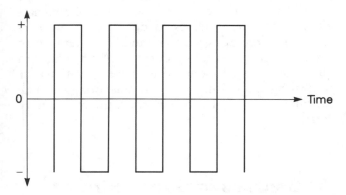

Figure 3.4 A graph of current or voltage against time for the circuit in Figure 3.3 if the switch is operated at regular intervals.

A p.d. that continuously reverses is known as an **alternating voltage**, and a current that continuously reverses is known as **alternating current**, abbreviation a.c.

The waveform shows that the transition from one direction to the other is more or less instantaneous. The waveform (which occurs very often in modern electronics) is called a **square wave** because of its appearance.

Suppose that the transition did not take place instantly. You saw in the previous chapter how it is possible to make a variable resistor, or potentiometer. Figure 3.5 shows, both as a drawing and as a circuit diagram, a modification of the standard variable resistor. This device (which is not a 'standard' electrical component but something we are imagining to help visualise what is happening) has a continuous resistive track, and two brushes, fixed together but insulated from one another. The brushes are each connected to one terminal of a battery. At opposite sides of the circular track, connections are made that enable the p.d. across the circle to be measured. Now let us look at this p.d. for various positions of the brushes.

Figure 3.6 shows the state of affairs if the brushes are immediately above the two fixed connections. It doesn't take a lot of thought to realise that in

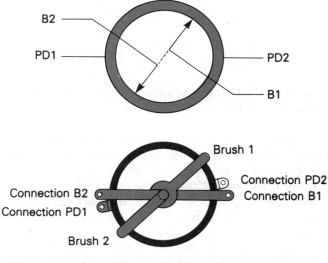

Figure 3.5 A modified variable resistor.

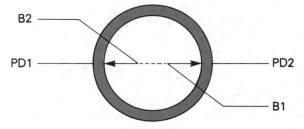

Figure 3.6 When the brushes are immediately above the two fixed connections.

these positions the measured p.d. will be that of the battery, either with one polarity or the other. Current will be flowing through both 'sides' of the circle, but this will not affect the results.

Now look at Figure 3.7(a), which shows the situation with the brushes 90° away from the fixed connections. What is the p.d. across the connections?

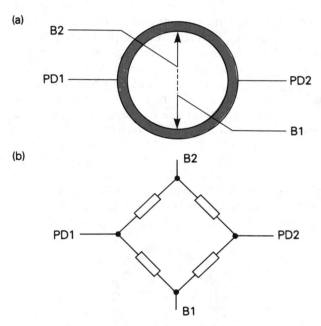

Figure 3.7 When the brushes are at right angles to the two fixed connections.

Figure 3.7(b) gives a clue to how this is calculated. Using arbitrary values for the four quadrants, Kirchhoff's laws will give the answer: the p.d. across the fixed connections is zero. Of course, the same thing applies if the brushes are in exactly the opposite positions, 180° further round the track.

So, starting from the 'zero' point, we rotate the brushes smoothly round the track. As we do this, the p.d. across the fixed terminals increases to its maximum positive value, drops back again through zero, increases to its maximum negative value, then back through zero after a full 360° rotation. Figure 3.8 shows one cycle of this waveform, which is known as a **sine wave**.

The length of time that it takes for the waveform to go through one complete cycle is called the **periodic time**. The usual way to measure alternating waveforms is to state their frequency; that is, the number of cycles they go through in one second. There is a special unit for 'cycles per second', called the **hertz** (abbreviation Hz), after

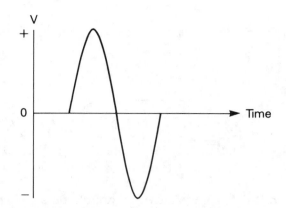

Figure 3.8 Graph of the p.d. across the fixed terminals of variable resistor in Figures 3.6 and 3.7.

Heinrich Rudolf Hertz, who carried out pioneering work on radio waves.

> Frequency can be calculated very easily from the periodic time using the formula:
>
> $$f = \frac{1}{T} \text{ hertz}$$
>
> where f is the frequency and T is the periodic time in seconds.

The rather strange potentiometer that I have described does not exist as a generator of alternating voltage, but it serves to demonstrate what is meant by alternating voltage (and, by extension of the same idea, alternating current). A little later in this book we shall be looking at 'real' systems for generating alternating current; but first we shall consider sources of direct current in rather more detail.

Sources of direct current

Primary cells

You may remember that I mentioned that the chemical effects of an electric current, demonstrated by electroplating and by electrolysis, can work both ways. A primary cell is a chemically powered generator of electricity, commonly known as a 'battery'. The word 'battery' – which my Con-

cise Oxford Dictionary says means (among other things) *'connected sets of similar equipment'* – has, by popular use, come to include 'cell'; technically, 'battery' should be used only to refer to a chemically powered generator of electricity that consists of more than one cell. So although I shall be talking about 'cells', these will be what most people call 'batteries'.

Leclanché cells

The simplest, and oldest type of cell is the Leclanché cell, named after its French inventor, Georges Leclanché, who patented it in 1866. A typical modern Leclanché cell is illustrated in Figure 3.9.

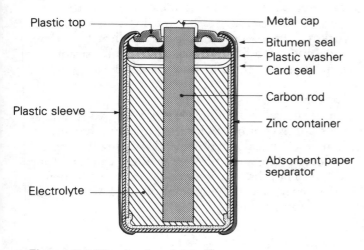

Plastic top — Metal cap
— Bitumen seal
— Plastic washer
— Card seal
— Carbon rod
Plastic sleeve — — Zinc container
— Absorbent paper separator
Electrolyte —

Figure 3.9 The Leclanché cell.

The modern version of the cell consists of a container, made of zinc, which is lined with absorbent paper. The inside of the cell is filled with powdered manganese dioxide, mixed with fine carbon paper (to make it conducting) and damped with ammonium chloride solution. A liquid or semi-liquid solution is used in almost all cells; it is known as the **electrolyte**. A carbon rod is used to make contact between this mixture and a terminal at the top of the cell. It is held in place by a washer, a seal made of bitumen (which prevents the water in the ammonium chloride solution from evaporating), and a plastic top cap. A steel jacket over the zinc case seals in the contents in case the zinc should corrode right through when the cell is exhausted.

Rather complex chemical reactions take place between the zinc casing, the ammonium chloride, and the manganese dioxide, the result of which is to cause an e.m.f. of between 1.6 and 1.2 volts to appear between the top cap (positive) known as

the **anode**, and the casing (negative) known as the **cathode**. When the top is connected to the casing by a conductor, an electric current flows through the conductor (Figure 3.10). Chemical reactions in the cell take place that maintain the e.m.f., for as long as the cell lasts.

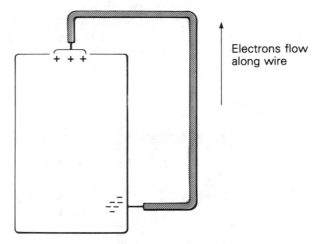

Electrons flow along wire

+ + +

Figure 3.10 Electrons flowing in a circuit with a cell. *Don't try this!*

Even if the conductor used has negligible resistance, the current will be limited by what is called the **internal resistance** of the cell. The internal resistance of a cell is very difficult to calculate; it varies widely between different types of cell, and the physical construction of the cell is an important factor. It can be estimated fairly accurately by measuring the e.m.f. of the cell and the amount of current flowing when it is short-circuited (that is, when the ammeter is connected directly between the two terminals of the cell). Special meters able to withstand high currents must be used for this measurement.

The internal resistance in ohms is given by

$$r = \frac{E}{I} - 0.01 \ \Omega$$

where r is the internal resistance, and E and I are the instantaneous e.m.f. and short-circuit current readings.

Manganese–alkaline cells

The manganese–alkaline cell is commonly available, under several brand-names, as the popular 'long-life battery'. The construction is more complicated than that of a Leclanché cell, although basically similar. It is shown, rather simplified, in Figure 3.11. The positive terminal at the top of the cell is connected to a dense layer, near the outside of the cell, consisting of compressed manganese dioxide and graphite. An absorbent separator cylinder is followed (working inwards towards the middle) by a paste of zinc mixed with potassium hydroxide. This is connected to the bottom of the cell by an internal post, riveted or welded to the bottom of the cell.

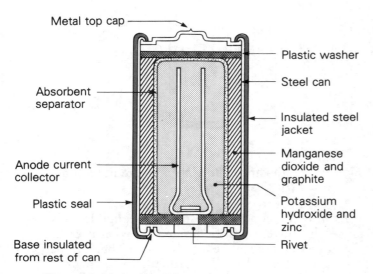

Figure 3.11 A simplified diagram of a manganese–alkaline cell.

The e.m.f. of a manganese alkaline cell varies between 1.5 and about 1.2 volts in use, but its capacity, or service life, is several times that of a Leclanché cell in most applications. Moreover, it has a much longer 'shelf life', and in normal conditions will retain 95 per cent of its original capacity after three years' storage; a Leclanché cell will be down to this level after less than a year.

Mercury cells
Because they contain relatively large amounts of mercury, which is a toxic metal, mercury cells are less widely used than they were. They were invented before the safer **silver oxide** cells (see below), and found uses in applications where a very stable source of e.m.f. is needed. The e.m.f. of a mercury cell typically varies between about 1.35 and 1.2 volts in use. Mercury cells are very like manganese–alkaline cells in construction, except that they are usually made as very short cylinders, or 'button cells'. The anode is made from high-purity zinc powder; the cathode is a mixture of mercuric oxide and graphite. They are separated by an absorbent disc soaked in potassium hydroxide, a very strong alkali. Mercury cells have a high capacity in relation to their weight, and storage properties that are similar to those of manganese alkaline cells. A disadvantage is cost: mercury is expensive as well as toxic.

Silver oxide cells
Similar in appearance to mercury cells, silver oxide cells provide a very stable and slightly higher e.m.f., at 1.5 volts. The cathode is made of silver oxide, with potassium or sodium hydroxide, both strong alkalis, as the electrolyte. The capacity of the silver oxide cell, size for size, is substantially better than that of the mercury cell. The high cost of silver makes them economic only for applications where size and stability of voltage is of paramount importance, such as in watches and hearing aids.

Zinc–air cells
These cells also look very similar to mercury cells but have about twice the capacity. They use air in their chemical reactions, and are usually supplied sealed; pulling a tab off energises the cell by letting air in. They have a very long storage life before being energised, an e.m.f. of about 1.4 volts, and are relatively cheap.

Lithium cells
Lithium cells offer the best energy-to-size ratio of any types discussed so far. They also have a much higher e.m.f., from 3.8 to 3.0 volts. Coupled with a very long storage life – 90 per cent capacity after five years – they are ideal for use as 'back-up' batteries for low-power computer memories, where they are often soldered into the circuits and have a life of several years.

Solar power
The sun is the energy source that powers the world's natural systems: plants, animals, the weather, everything. In full sunshine, around one kilowatt falls on each square metre of ground, so it is sensible to use this power if we can, particularly in countries where there is hot sun and not much cloud.

There are two basic ways of using solar power to make electricity. The first is to use the heat of the sun to drive a motor of some kind, which in turn operates a generator. The second way is to convert the light directly into electricity using **solar cells**.

Because solar cells are expensive for the amount of power they produce they are suitable for powering devices that use relatively little current, or in which cost is unimportant. Details of the way in which solar cells (or photovoltaic cells to give them their proper name) work must be left until after a study of semiconductors. However, they are surprisingly efficient. Small solar cells are often used to power pocket calculators, and sometimes to charge batteries when there are no other power sources nearby. Big arrays of solar cells are very expensive, and at present are not practical for generating domestic or industrial power. For specialized applications – particularly where only small amounts of energy are needed (for example, calculators) or where sunlight is plentiful and cost is no object (space vehicles and satellites) – solar cells are often ideal.

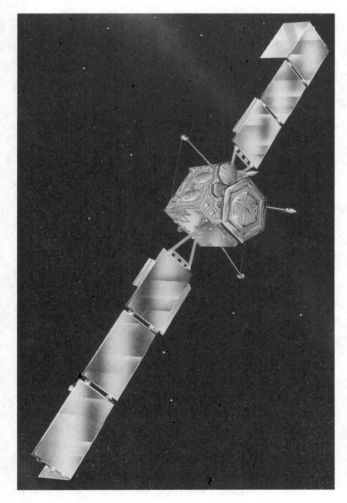

Figure 3.12 A high-technology example of solar power in use: the *INMARSAT-2* communications satellite. (Photograph courtesy of British Aerospace)

Secondary cells

All the above cells provide a source of e.m.f. while the chemicals last. When the cell is exhausted, it has to be thrown away. Secondary cells (also known as accumulators) can be **recharged** once they are exhausted, by connecting them to a suitable d.c. supply. This makes them more economical in the long term. There are two types of secondary cell in common use.

Lead–acid cells

The most common of all secondary cells is the lead–acid cell, used in car batteries. A typical lead–acid cell is shown in Figure 3.13. The positive and negative plates are both made of lead (or a lead alloy) and are shaped like a waffle, containing many small square holes. The holes in the positive plates (the cathode, you will remember) are packed with lead peroxide, and the holes in the negative plates (the anode) are packed with spongy lead, which is lead treated to give it the maximum surface area. The electrolyte is dilute (but still strong) sulphuric acid.

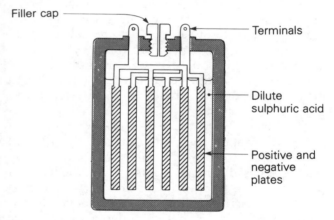

Figure 3.13 A lead–acid cell.

Lead–acid cells have an e.m.f. of about 2 volts, and a very low internal resistance. This allows extremely high currents to be supplied for brief periods. A car battery, if shorted between its terminals, can provide a current running into hundreds of amps. This is enough to do a lot of damage to the battery or to whatever is shorting it. I have seen a screwdriver, dropped between the terminals of a car battery, blown in half. It always pays to take great care when dealing with lead–acid accumulators; if shorted, the worst that can happen is that the battery will explode, showering corrosive sulphuric acid in all directions.

> ⚠ The sulphuric acid in a car battery is very corrosive, and if you mop it up with a rag and then put the rag in your jeans pocket, it will rapidly eat a hole in both the rag and your jeans. Acid can also burn your skin, so be very careful when handling lead–acid batteries.

When a lead–acid cell is fully discharged – that is, no more current can be taken from it – both the spongy lead and the lead peroxide have been converted into lead sulphate. Some of the sulphuric acid has been converted into water, which makes the electrolyte less dense. It is therefore possible to determine the state of the charge of a lead–acid cell by measuring the relative density of the electrolyte. For a fully charged cell, the density would be about 1.26, dropping to about 1.15 when fully charged.

To recharge the cell, it need only be connected to a d.c. supply, at a voltage that will ensure that the cell is fully charged within about 10 hours. Towards the end of this time, oxygen will bubble off the positive plates and hydrogen will bubble off the negative plates. (You ought to be able to guess why this is. If you can't, then re-read the part of Chapter 2 of this book that deals with the chemical effects of an electric current.) It is necessary to make provision for these potentially explosive gases to disperse.

The advantages of the lead–acid cell are that it is cheap to produce, has a good capacity for its size, and is able to provide very large currents if required. Disadvantages are its weight, the fact that the electrolyte is both liquid and very corrosive, and the necessity to allow gas to escape while it is being charged. If you have wondered why cars are often reluctant to start in cold weather, part of the reason is the fact that lead–acid cells lose a lot of their efficiency in cold weather, and in sub-zero temperatures the capacity is sharply reduced.

Nickel–cadmium cells

The nickel–cadmium cell was patented in 1901 by a Swede called Waldemar Jugner, but it was not until the 1950s that a sealed nickel–cadmium cell became a practical proposition. A typical cylindrical nickel-cadmium cell is shown in Figure 3.14, although 'button' types are also commonly available.

The internal structure of the cell is a roll, based on a perforated nickel strip. The positive electrode

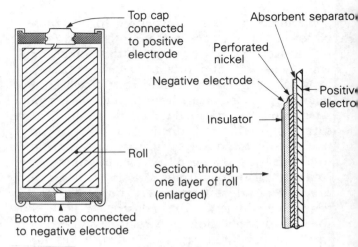

Figure 3.14 A typical nickel–cadmium cell.

(cathode) consists of microporous nickel, supported by the perforated strip. The negative electrode is microporous cadmium, or a mixture of cadmium and iron. A strongly alkaline electrolyte is used. The e.m.f. resulting is relatively constant at around 1.2 volts.

The advantages of a nickel–cadmium cell are considerable. It can be produced as a sealed unit, so it can be used any way up, and in portable equipment. It is capable of very large discharge currents (even more than the lead–acid cell). It is relatively light, has a good capacity-to-volume ratio, and can withstand modest overcharging indefinitely. It works better than a lead–acid cell battery at low temperatures (although the capacity is still reduced when the cell is cold). It requires no maintenance.

Disadvantages are that a nickel–cadmium cell will not retain its charge very long; over half its charge will have leaked away in three months. It gets hot when being charged at fast rates. Finally, cadmium is a very toxic substance and must be handled carefully.

Other secondary cells

Nickel–iron cells (known as 'NiFe batteries' after the chemical symbols for nickel and iron) used to be popular and can still be found in place of lead–acid batteries in some places. Although cheaper and lighter than lead–acid cells, they produce a lower e.m.f. at around 1.2 volts, and have a worse capacity-to-volume ratio. Work is continuing on the development of more efficient and cheaper storage cells, but so far nothing seems set to displace the lead–acid accumulator for high-power applications, or nickel–cadmium for applications in electronics.

Cell capacity

The capacity of a primary or a secondary cell is measured in **ampere-hours**, abbreviation Ah: that is, the number of hours for which the cell can sustain a discharge of 1 ampere. This figure can be misleading, as it is not meant to imply that the cell in question is necessarily capable of supplying a current of 1 amp.

A small nickel–cadmium cell ('NiCd' is often used as an abbreviation) might be able to supply a maximum current of only 250 mA. Such a cell would be rated at 1 Ah if it could keep up a 250 mA discharge for 4 hours. The capacity of a cell (primary or secondary) changes according to the rate at which it is discharged, and in many types of cell the capacity is increased if the cell is allowed to 'rest' in between periods of discharging. To quote a cell's capacity in ampere-hours is nevertheless a good guide.

Charge and discharge cycles

When a secondary cell or battery is charged, it is always necessary to put in more than you take out. The proportion of power that has to be put in, compared with what can be taken out again, is called the charging factor. For a lead–acid car battery, the charging factor is about 1.3. For example, you would have to charge a completely discharged 20 Ah battery for about 20 × 1.3 = 26 hours at 1 ampere to charge it fully. Nickel–cadmium cells have a charging factor of about 1.4.

Batteries

A battery, as I mentioned above, is actually a set of cells connected together in **series**. Batteries almost always consist of several cells in the same container, usually designed with the object of increasing the e.m.f. (voltage). When cells are connected together in series, as in Figure 3.15, the e.m.f. of each cell is added together.

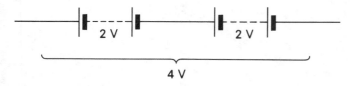

Figure 3.15 A battery of cells.

Thus six lead–acid cells, e.m.f. 2 V each, are connected together in a car battery to provide 6 × 2 = 12 V for the car electrical system. Any cells,

either primary or secondary, may be connected together in series to give an increased voltage. It is important to realise that the internal resistances of each cell are also added together, so the absolute maximum current capability of six cells will be only one sixth that of a single cell. This is seldom important when connecting together reasonable numbers of secondary cells in series, but may be significant when connecting several primary cells (especially Leclanché cells) together.

Most cells can be connected together in **parallel** to increase the current capability and capacity without altering the e.m.f. The internal resistance of a parallel combination of two identical cells is half that of the individual cell; the capacity is doubled. Nickel–cadmium cells must not be connected in parallel. If you do, one cell is likely to discharge into the other, probably ruining them both.

Source of alternating current: the electric generator

We saw at the beginning of this chapter that electric current can be **direct** or **alternating**. It is not possible to obtain anything other than d.c. from a primary, secondary, or solar cell, but alternating voltages and currents can be produced by other means, typically **electric generators**. To understand how generators work, we must first of all look at the relationship between magnetism and electricity.

You have already seen how electric current produces magnetism (and I hope you have carried out an experiment to demonstrate it). It is equally possible to demonstrate how you can produce electricity using a magnet. A very simple experiment involves nothing more complicated than a bar magnet, a coil of wire, and a sensitive voltmeter. Figure 3.16 shows how the experiment should be performed. The coil should have as many turns as possible on it. An aerial coil from an old transistor radio is quite suitable, particularly if you can find a bar magnet that will fit through the middle of it. Otherwise you could wind the coil yourself using thin **insulated** wire, preferably enamel-coated coil wire. You will need at least 200 turns on the coil.

There are a couple of points that are worth emphasising in this experiment. First, the meter indicates that electricity is produced only when the magnet is *moving* relative to the coil. Second, *if*

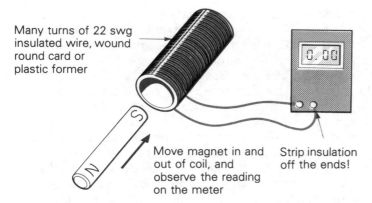

Many turns of 22 swg insulated wire, wound round card or plastic former

Move magnet in and out of coil, and observe the reading on the meter

Strip insulation off the ends!

Figure 3.16 *An experiment to produce electricity using a coil of wire and a magnet.

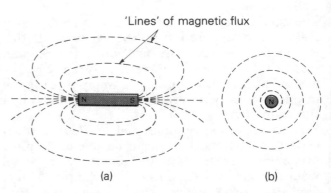

'Lines' of magnetic flux

(a)   (b)

Figure 3.17 Lines of magnetic flux round a bar magnet.

the magnet is reversed, the electric voltage is of the opposite polarity, and *the current flow is reversed*.

Magnetism and models of reality

At this point, it might occur to someone to ask the question, 'What is magnetism?' Like many questions in physics, there is no definite answer to this question. We can describe magnetism's effects very accurately, but this is not the same as describing magnetism itself. We can define rules about magnetic behaviour, but as yet we cannot say *why* magnetism behaves the way it does with any degree of conviction.

Physicists can describe magnetism mathematically. However, a mathematical description of – for example – a tennis ball and the way it bounces isn't the same thing as saying what a tennis ball *is*. It is always important in science to recognise the difference between **describing** and **understanding**.

We speak of a magnet having a **magnetic field** associated with it. We can map the magnetic field, measure its strength, and describe its properties, all without really knowing what the field actually is. If we cannot visualise (or in some cases fully understand) what is happening, we can at least make a mental **model** of it, that shares some of the properties of the real thing. Fortunately, this is all we need to do; the bow and arrow were invented long before anyone knew what it is that makes wood springy!

Try this quick experiment. Take a bar magnet, place a sheet of paper over it, and sprinkle iron filings on the sheet of paper. The iron filings will make a pattern that connects lines of equal magnetic strength, 'flowing' (although magnetism does not really 'flow' anywhere) between the magnet's

poles. It looks like Figure 3.17(a). For convenience we *pretend* that the magnetism flows from the north pole of a magnet to the south pole, and we call the imaginary field lines **lines of magnetic flux**. The word 'flux' means 'flow', but remember that this is only a model, and nothing really flows at all.

We can also plot the field looking at the end of the bar magnet; it appears to be 'flowing' around the edge of the magnet in one direction for the north pole, and in the other direction for the south pole (Figure 3.17(b)).

Now imagine that, in place of the magnet, there is a length of wire. Clearly, since the ends of the wire are inaccessible, it is hard to measure the magnetic field looking at the length of it (as we did in Figure 3.17(a)). We can, however, look at the field surrounding the wire, and it turns out that it is identical to that shown in Figure 3.17(b).

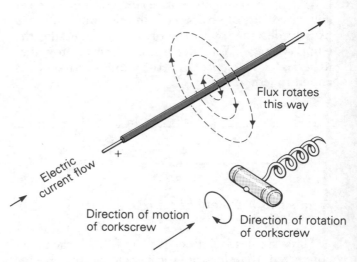

Flux rotates this way

Electric current flow

Direction of motion of corkscrew

Direction of rotation of corkscrew

Figure 3.18 The corkscrew rule for magnetic flux.

Figure 3.18 shows how we can remember the direction of 'flow' of the magnetic flux. This is called the **corkscrew rule**, presumably because the European scientists who did the pioneering work on electricity were all very familiar with opening bottles of wine with corkscrews! We might equally call it the 'woodscrew' rule: current flowing in the direction of the screw's movement causes a clockwise flow of magnetic flux.

Remember that almost all of this is imaginary. It is a **model** for magnetic behaviour, useful in designing electrical equipment; nobody is suggesting that it is real.

While we are on the subject of what we do and don't know, I shall take the opportunity to mention an important fact that electronics engineers, secretly, are a little ashamed of. It is always assumed that electric current flows from the positive terminal of a source of e.m.f. to the negative terminal. You will often find little arrowhead symbols used in circuit diagram components to indicate the direction of current flow. Electric current, we say, flows from positive to negative.

Actually, it doesn't. Electrons move the other way, from negative to positive, opposite to what has been called 'conventional current'. This is an odd fact, but one that rarely complicates circuit design and analysis. The reason for it, frankly, is that someone made a big mistake.

When the science of electricity was in its infancy, there was a need to standardise (in books and notes) on the direction of current flow. In the absence of our knowledge of electrons and atoms, someone sat down to perform an experiment that would settle the problems of current flow once and for all. It might have been Volta, or Faraday, I can't remember which. The experiment consisted of setting up equipment that would make a very long electric spark, and then repeatedly watching the spark. Eventually the experimenter made up his mind that the spark always jumped from *here* to *here*, and unfortunately he got it wrong.

When, years later, it was discovered that (1) electrons go from negative to positive, and (2) electric sparks jump much too fast to see which way they are going anyway, 'conventional current' was established in all the textbooks and it would have caused hopeless confusion to try to change it.

So we still say that 'conventional current' flows from positive to negative, and pretend that it does so when drawing circuit diagrams. The only time we use 'electron current' is when we are discussing semiconductor physics, where it really matters. Back to the magnets.

Solenoids and electromagnets

If we take our current-carrying wire and bend it round into a circle, it produces a field pattern like the one shown in the perspective drawing in Figure 3.19. You can see that the middle part of this field looks just like the field produced by the bar magnet; look back at Figure 3.17(a). The loop of wire does, in fact, behave just like a bar magnet. We can make the magnetism more powerful by adding more loops, winding lots of turns of wire into a coil. When the current flows, the magnetism produced is indistinguishable from the field of an equivalent bar magnet. The only difference is that we can turn the **electromagnet**, as it is called, on and off. An electromagnet consisting of a helical coil of wire is known as a **solenoid**.

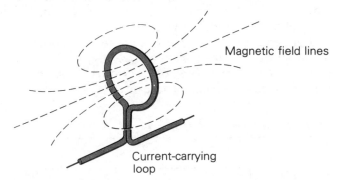

Magnetic field lines

Current-carrying loop

Figure 3.19 Magnetic field round a loop of wire.

There is an easy way to remember which end of the solenoid is the north pole and which is the south pole. The letters 'N' and 'S', if drawn on the ends of the solenoid, point the way the current is flowing (Figure 3.20).

Figure 3.20 An easy way to remember the polarity of the ends of a solenoid.

We have been considering a solenoid consisting only of a number of turns of insulated wire. If we add to this a **core** made of iron, the strength of the magnetic field produced will be much greater. The magnetic force seems to be concentrated by the iron core. Various materials can be used in the core of an electromagnet, most of them based on iron. They are known as **ferromagnetic materials**. You can test for yourself whether a metal is ferromagnetic; if it sticks to a permanent magnet, it is ferromagnetic.

Some ferromagnetic materials, such as pure iron, are said to be magnetically 'soft'. Such materials do not retain much (if any) magnetism when the electric current flowing through the solenoid is turned off. Iron is therefore a good metal to use in making an electromagnet that is used to move something or pick something up. Other ferromagnetic materials are magnetically 'hard': that is, they retain most of their magnetism when the power is switched off. It is possible to make iron magnetically hard (not physically hard, like steel) by adding small amounts of aluminium, nickel, cobalt and copper to it. 'Hard' magnetic materials are useful for making permanent magnets.

Magnetically 'soft' materials are more important in electrical and electronic engineering, where they can be used for the cores of electromagnets, transformers (see below), and other items.

Electromagnetic induction

> If a conductor (such as a wire or coil) is in a magnetic field, then any changes in the strength of the field will cause an e.m.f. to be induced in the conductor.

You should remember this fundamental rule of magnetic induction.

It is essential for the field strength to *change* for induction to occur. A field of constant strength will not induce an e.m.f.

Earlier, I suggested that you try an experiment in which a magnet was passed through a coil to cause an induced e.m.f. – and thus an induced current – to register on a meter. The field of a permanent magnet does not change, so how did this happen? You may already have guessed that **moving the magnet** was what caused the induced e.m.f. The motion imparted to the magnet caused the field 'seen' by the coil to increase and then decrease again. As far as the coil was concerned the field was changing in strength.

When we plot the lines of a magnetic field (as in Figure 3.17), we join together points of equal field intensity, in just the same way that contours on a map join together points of equal height. If we were to take a conductor and move it carefully along one of these lines in the field of a powerful magnet, no e.m.f. would be induced because there is no change in the field experienced by the wire. On the other hand, moving the wire *across* the lines of force changes the field strength and induces an e.m.f. in the wire. This principle is known as **electromagnetic induction by motion**, and is the principle on which the electric generator works. Figure 3.21 shows a sketch of the very simplest possible electric generator, but you will no doubt realise that this generator is more theoretical than practical!

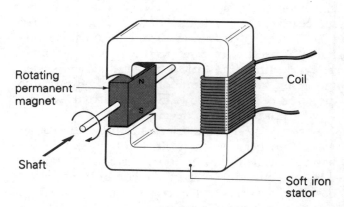

Figure 3.21 A very simple electric generator.

As the magnet is rotated, the field 'seen' by the coil is constantly changing. During the course of one rotation it increases to a maximum value, decreases to zero, reverses, then increases to a maximum value in the other direction.

Does this sound familiar?

The induced e.m.f. follows the sine-wave pattern that we saw at the beginning of this chapter. The mains electricity supply is a sine-wave alternating current, because it is generated in just this way.

In real generating stations it is usual to have a generator in which the magnet stands still and the coil spins round, but there is no difference in the principle; it is just that the coil can be made a lot lighter and easier to spin than the magnet.

The source of alternating current for power supplies is generators. As we shall see in the next chapter, various power sources can be used to spin the moving parts of the generator, but electrically they all work in more or less the same way, from a bicycle dynamo to a hydroelectric power station.

The measurement of alternating current

Because steady-state alternating currents – sine waves such as mains electricity – are so common in electrical engineering (and to a lesser extent in electronics), it is worth looking at the ways in which such changing values are measured. You should remember from my earlier mention of alternating voltages that the sine wave's **periodic time** and its **frequency** tell us about the way it changes with time.

The voltage of an a.c. source is continuously changing. The voltage (or current) at any given instant in time is called the **instantaneous value**, and the highest value it attains (positive or negative) is called the **peak value**, or sometimes **amplitude**. The peak value is fairly easy to measure, as you will discover in the next chapter.

It is reasonable to assume that a given alternating current passing through a load (say, a resistor) will dissipate a given amount of current. You have seen in this chapter that it is an easy calculation to make for direct current.

When considering the power dissipated by a component through which a.c. is flowing, we use a measurement called the **rms value** (pronounced 'are-em-ess'). In this context 'rms' stands for 'root mean square', which refers to the way in which the value is calculated.

> The rms value of an alternating current is the equivalent value to the direct current that would dissipate the same amount of power when passed through a resistor.

If I_p is the peak voltage attained by the sine wave (its amplitude), then the rms current (I) is equal to

$$I = \frac{I_p}{\sqrt{2}}$$

$\sqrt{2}$ is the square root of 2, which is about 1.414, so

$$I = \frac{I_p}{1.414}$$

In practice, it is easier to state this as

$$I = I_p \times 0.707$$

which means that the rms value of an a.c. sine wave is 0.707 times its peak voltage. When people refer to 'the mains voltage' being either 220 V or 110 V, they are referring to the rms value.

Static electricity

We need to look at one more – rather odd – form of electricity. This is **static electricity**. We have so far considered only moving, or dynamic electricity. But it is possible for an electric charge to exist at rest. If a charge can be induced in an insulating material, then the charge will be unable to flow away through the material. But how can an electric charge be imparted to an insulator?

It is actually quite easy to do. A hard plastic rod and a silk or nylon cloth are all that you need. First make sure that both the cloth and the plastic are completely dry. Then hold the end of the plastic rod and rub the rod vigorously with the cloth. After a short time, the plastic rod will become charged. The electric field is so intense that it is possible to pick up tiny fragments of paper by electrical attraction.

This demonstration will work best in a warm, dry climate. If you live in a country where the atmosphere is damp (or if it is raining when you try the experiment) you may not be able to get a good result, because water in the air conducts away the electric charge. The voltage present in static electricity is very high indeed, although the current available is very tiny. In a dry climate, it is possible for a person wearing plastic- or rubber-soled shoes to become charged up to several kilovolts! This isn't dangerous – because the current is so small – but it can lead to an unpleasant shock when you touch something that conducts the charge away. Pulling off a nylon sweater in the dark in a warm, dry climate will often result in a visible display of tiny electric sparks.

Static electricity is important in electrical and electronic engineering for a number of reasons. Here are two of them.

Some electronic components are unable to withstand high voltages. Certain types of integrated circuit (microelectronic chips) can be completely ruined by electrical potentials of more than a few tens of volts, regardless of whether or not dynamic or static voltages are involved. In other words, just touching the connections of such a circuit could destroy it! We therefore have to be careful about handling such devices, and take special precautions when handling them.

A special kind of static electricity is **lightning**. During storms, **thunder cells** can develop, in which a weather system becomes charged up to an enormously high potential relative to the ground. If the p.d. between the thunder cell and the ground is large enough the electricity will discharge, causing a very large spark to jump between the cloud and the ground: lightning.

The amount of energy involved in lightning is tremendous. The p.d. might be as high as a thousand million volts (1 GV), with a current of more than a million amperes (1 MA). The energy released can knock down buildings and start fires. Lightning is important in that it is inevitable that lightning will strike power distribution and telephone lines from time to time. Various safeguards have to be built into the power distribution and telephone networks to minimise the danger caused by lightning.

Buildings can be protected to some extent by the use of lightning conductors. The lightning conductor consists of a sharply pointed rod, mounted on the highest part of a building and connected to the ground by a heavy metal strap, running down the outside of the building. A lightning conductor works in two ways. First, it conducts away some of the electric field in the air in the vicinity of the rod, which reduces the likelihood of lightning striking that particular building. Second, if lightning does hit the building, it gives the discharge an easy path to the ground, along the outside of the building, so that less damage is caused.

■ CHECK YOUR UNDERSTANDING

● Electric current can be **direct** (flowing in one direction) or **alternating** (flowing backwards and forwards).

● The frequency of alternation of alternating current (a.c.) is measured in **hertz** (symbol Hz). One hertz is the same as one alternation per second.

● **Primary cells** provide a source of e.m.f.; they are powered by chemicals. Primary cells cannot be recharged.

● **Secondary cells** also provide a source of e.m.f., but they can be recharged. When charging a secondary cell, you always have to put in more than you get out; some energy is wasted.

● **Induction** is the name given to an aspect of the magnetic effect of an electric current, in which a conductor moving through a magnetic field becomes a source of e.m.f.

● **Generators** convert mechanical energy into electrical energy by using induction.

● **Static electricity** is the name given to an electric charge in an insulating material. Electrostatic voltages are often very high and can damage delicate electronic components.

REVISION EXERCISES AND QUESTIONS

1 What is the difference between direct and alternating current? What units are used to measure alternating current?
2 What is the different between a primary cell and a secondary cell? Give an example of each.
3 A battery is marked '4.8 V 3 A'. Why is is unlikely that this is a Leclanché battery? Make a guess about what kind of battery it might be, and give your reasons.
4 What is the requirement for an electic current to be induced in a wire that is placed in a magnetic field?
5 What is the r.m.s. value of a sine-wave alternating current? What relationship does it bear to the peak current?
6 What is meant by 'conventional current'?

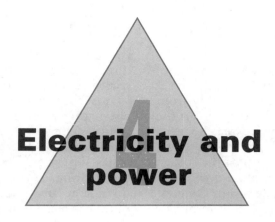

Electricity and power

Introduction

In this chapter we shall examine the ways in which electricity is generated in practice. Various different fuels can be used, and all of them appear to have both advantages and disadvantages. Comparing them is interesting and gives an insight into the problems that have to be solved in getting an electricity supply to everybody who wants electric power. At the end of the chapter I shall talk about electric motors: the other end of the power line, so to speak.

The creation and generation of electric power

Generators

You read, in the previous chapter, how – in principle – an electric generator works. A coil of insulated wire, spun in a strong magnetic field, will cause an alternating e.m.f. in the wire, generating an alternating current. In almost all power stations, the electric generators use this principle.

A full-size generator is divided into two parts: the **stator**, or fixed part, which produces the magnetic field in which the **rotor**, or rotating part, is spun. Because it is impractical to provide gigantic permanent magnets to create the required magnetic field, it is usual for a large electricity generator to use electromagnets to produce the field.

This means that the generator has two electrical components. One part produces direct current for the **field windings** (the coils that produce the magnetic field). The other part produces alternating current (the generator's output), which is fed into the public electricity-distribution network.

Because direct current is needed to make a steady electromagnetic field, the part of the generator that powers the field windings has to produce direct current. A piece of machinery called a **commutator** is used to do this. A simplified generator with a commutator is illustrated in Figure 4.1.

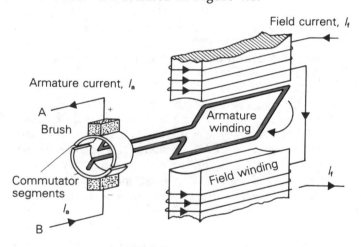

Figure 4.1 A simplified direct current generator.

On the right-hand side of Figure 4.1 is the **field winding**, which is supplied with a field current (I_f) to produce an electromagnetic field. The coil, which is properly called the **armature winding** (just a single turn in this case, to make the drawing clearer) rotates in this field. On the left of Figure 4.1 is the **commutator**. This consists of two conducting half-rings called **commutator segments**, each of which is connected to one end of the armature windings. A pair of conducting **brushes** carry current from the commutator segments.

When the generator is working, an e.m.f. is induced in the armature winding, and is available at the output terminals marked A and B in Figure 4.1. You will no doubt remember from Chapter 3 that the output is a **sine wave**, rising steadily from nothing to a peak, then down to zero again. At the precise moment that the sine wave reaches zero the commutator segments reach the point where they move between one brush and the other, reversing the connections to the armature winding. When the sine wave begins to rise again, the connections between the coil and the output terminals have been swapped over.

The output across A–B therefore looks like the waveform shown in Figure 4.2; the e.m.f. is always of the same polarity, and so the output is a direct current. The current is not continuous, but the flow is always in the same direction.

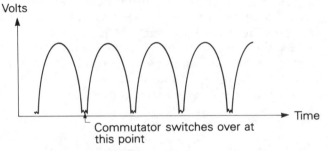

Figure 4.2 Graph of the voltage produced by a simple d.c. generator.

A full-sized generator has not just one, but many coils, wound so that each is a few degrees further round the rotating armature. Each coil has its own pair of commutator segments. The combined output produces a direct current that, while not completely smooth, is at least more or less constant. A graph of the output of a generator with three windings is shown in Figure 4.3.

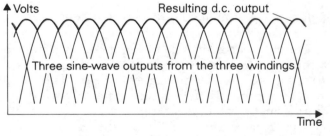

Figure 4.3 Graph of the voltage produced by a three-winding generator.

If a direct current output is required, then a generator built according to this design would be all that is needed. Unfortunately – and for reasons we shall look into a little later in this chapter –

direct current is not suitable for the public electricity supply. Alternating current is essential.

A second part of the generator produces the alternating voltage that is fed into the public electricity distribution network. It is rather simpler than the d.c. generator. In place of a commutator, it has **slip rings**. The slip rings carry current from the rotating armature to the output terminals, but do not at any stage reverse the polarity, so the output is a.c. A simplified sketch of an a.c. generator, known as an **alternator**, is given in Figure 4.4.

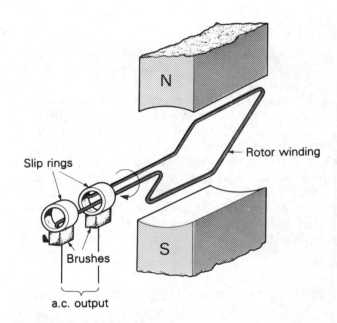

Figure 4.4 An alternator.

In a full-scale public electricity generator (most people call them 'generators', although you now know they are really 'alternators') the magnetic field is provided, as we have just seen, by the field windings that are powered by the d.c. generating coils. In small (for example, bicycle) alternators, permanent magnets are used.

The frequency of the sine wave output depends on the speed of rotation of the rotor in the generator. In most African countries, as well as in the UK and Europe, the frequency of the mains supply is 50 Hz. This means that the rotors of the generators need to spin at the rate of 3000 revolutions per minute.

When the generator is 'off-loaded', the amount of resistance to spinning the rotor is relatively little. But as more and more current is taken, magnetic effects act to oppose the rotation of the rotor, and more power is needed to turn it. The amount of power required to work the generator therefore depends on the electrical load.

Single-phase and three-phase systems

The generator considered above is a single-phase generator; that is, the output is an alternating voltage that is produced by a single armature winding. Practical generators can be made more efficient by using three separate armature windings, spaced so that that the peaks of the three sine-wave outputs are 120° apart. The three windings are connected together through three terminals, so only three slip rings are required. Figure 4.5 shows the way the armature windings are usually connected, in what is called a **delta connection**. Each winding is capable of supplying current to a load, so it is possible to connect a load between any two of the three output terminals.

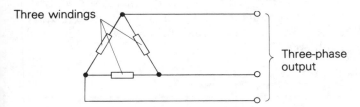

Figure 4.5 Delta connection in a three-phase generator.

It is important to realise that one effect of the windings being placed at 120° in relation to each other is to prevent current from flowing round the three interconnected windings; if this were not the case, current would flow round the armature windings and be lost – ultimately – as heat. A system with three windings (the most commonly used) is called a **three-phase** system. A graph of the output voltages of a three-phase system, superimposed on each other, shows the relationship of the three sine waves (Figure 4.6).

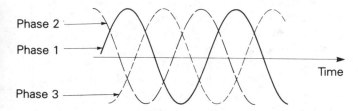

Figure 4.6 Output voltages from the three-phase generator superimposed on each other.

Three-phase power systems can be used in one of two ways. Major users of electricity, like factor-

ies, can use equipment that involves all three phases; a load that uses all three phases equally is called a **balanced load**. Three-phase electric motors can be made that are smaller and more efficient than single-phase motors. Motors are discussed later in this chapter. Alternatively, the phases can be used separately. A load that makes use of only one phase is called an **unbalanced load**. Domestic electric systems are of this type.

Although national power supplies use the three-phase system, which represents the most cost-effective solution to power distribution, don't forget that there is nothing special about using three phases. A generator can be made with any number of armature windings, and – for example – a five-phase system is technically quite possible; it just doesn't make commercial sense.

Sources of energy

What turns the generators in a power station? In all but hydroelectric plants the answer, perhaps surprisingly, is **steam**. A boiler full of water is heated to produce steam, and the steam, under very high pressure, spins a turbine. The turbine is connected by a shaft to the generator. The water in the boiler can be heated by means of one of a number of fuels.

Fossil fuels

Coal is used extensively in power stations throughout the world. Coal-fired power stations are often built near coalmines, or near ports where imported coal can be brought directly to the power station. **Oil** is widely used. This can be piped directly from the refinery to the power station, or shipped in oil tankers by sea. **Natural gas**, if it is available, is often used in preference to coal or oil, mainly because gas-fired boilers are cheaper to make.

All three of the above are what is known as **fossil fuels**, because they are produced, deep under the earth, from the remains of ancient animals and plants. Using them as fuel, while attractive in the short term, has two serious long-term disadvantages.

First – and most obvious – all three of these fuels will eventually run out. No more coal, oil or gas is being produced under the earth today, as the lifeforms of which they are composed are long-since extinct. Less than 200 years after the industrial revolution, we are already worrying about coal, oil and gas reserves. It seems likely that in less than 200 years mankind will have used the natural resources that it has taken nature millions of years to make.

Second, all three of these fuels release carbon dioxide when they are burned. Carbon dioxide in the atmosphere is taken up by all plants and animals, and over the millennia many billions of tons of it have been absorbed and then locked up under the ground in the form of oil and coal. This may have changed the earth's atmosphere significantly since prehistoric times. When we burn fossil fuels, we put the carbon dioxide back into the atmosphere, and many scientists fear that this may result in the world getting hotter: the so-called 'global warming' effect.

Pollution from fossil-fuelled power stations may also be causing serious damage to forests; high concentrations of carbon dioxide around power stations mix with rainwater to form **carbonic acid**, which falls on the forests as what is called 'acid rain'.

Atomic energy

Atomic energy is used in many countries to heat water in a power station's boilers. Radioactive **uranium** is used in an **atomic reactor** to produce heat, by a process called **atomic fission**. Atoms of the uranium break down and release a great deal of heat, which can be used to make steam and drive a turbine. Uranium is mined in many countries, but the radioactive part must be separated out and purified before it can be used for atomic power.

Another type of atomic reactor – called a **fast breeder** reactor – used a radioactive metal called **plutonium**. Plutonium is not found naturally, and is extremely dangerous. A tiny speck of plutonium can kill you, causing cancer or one of several other fatal diseases.

Atomic fission power has both advantages and disadvantages. The advantages are:

1. Unlike supplies of coal and oil, the supply of fissionable materials is unlikely to run out in the foreseeable future.
2. Unlike coal- and oil-fired stations, atomic power stations put no chemical pollution into the atmosphere and will not contribute to global warming.

There are also serious disadvantages:

1. Atomic power stations have a potential for disaster on a massive scale, demonstrated by the 'incident' in Chernobyl in Russia in 1986, which killed many people and has greatly increased the incidence of fatal cancers in the region. A big area around Chernobyl may always be uninhabitable.
2. The highly radioactive waste products of atomic power stations are difficult to dispose of safely, because they are poisonous and stay radioactive for thousands of years. Can you think of anywhere you could put a drum of deadly waste so you could be sure it would stay safe and intact until the year AD 4000?

A 'high tech' solution that is being researched (but will not be ready for many years) may be **atomic fusion**, which in theory at least should provide the advantages of atomic power in far greater safety and without the difficult waste disposal problem.

Hydroelectric power

Hydroelectric power uses the energy of falling water to generate electricity. It is the best of all power sources, and has almost no disadvantages: there is no pollution, and it doesn't use fuel. In fact its only major disadvantages are that it relies on naturally occurring waterfalls that flow all year round, or on rivers that can be dammed to make a waterfall. If an artificial lake has to be created to supply a hydroelectric power station, there may be all sorts of environmental problems. There may be benefits, too.

In Nigeria, for example, the Kainji dam on the river Niger allowed a hydroelectric power station to be built, but tens of thousands of people had to be evacuated from their homes on the proposed site of the lake. However, the dam provides a useful road across the Niger, provides water for irrigation, and reduces the risk of flooding down-river.

Hydroelectric power is based around a dam, designed to let water flow down through a huge pipe called a **penstock**. As the water falls through the pipe, it picks up speed (and energy) until it passes through a turbine at the bottom. It flows out of the turbine along the **tail race**, where it loses speed and rejoins the original watercourse. The spinning turbine drives a generator, which produces electricity.

Hydroelectric **wave power** is being researched.

No really satisfactory method of trapping the power of waves in coastal seas has yet been devised, but several are under active development.

Solar energy

Solar power lends itself to small installations, probably better than it does to large ones. In countries where there is a lot of sunshine, solar power is available, free, for the cost of collecting it.

The key part of any solar energy system is the **collector**. The simplest kind of collector is really no more than a heat-absorbing container (such as a black plastic tube or bag) filled with water and exposed to sunlight. The sun's heat is absorbed by the water. The amount of heat that is collected in this way can be quite considerable, but it is low-grade energy; the water is warm rather than hot and cannot be used directly to power a boiler or a motor.

Hotter water can be obtained by using a means of concentrating the sun's heat on the absorber. A reasonably efficient – and quite cost-effective – method is to use a **line-axis concentrating collector**, such as the one shown in Figure 4.7. The absorber is a black pipe, along which water is pumped. The sun's rays are collected and concentrated by a polished metal reflector, curved into the shape of a parabola. The design is known as a **parabolic trough concentrator**; it has the advantage that it is easy to construct and quite efficient.

Both the designs shown above are 'low-technology', being both simple and fairly cheap. They have the disadvantage of producing low-grade energy, and paradoxically a 'high-technology'

device – a **heat pump** – is needed to produce the kind of energy that is needed before they can generate electricity.

For large-scale solar power (small power stations) it is possible to use a much bigger and more efficient reflector system to produce high-grade energy in the form of steam. A very large reflector is expensive and unwieldy, so many smaller mirrors are set up in a field, carefully angled to reflect sunlight onto an absorber, which is usually mounted in a tower. With so much sunlight focused on it, water in the absorber boils almost instantly, and the steam that this produces can drive a turbine that works an electric generator. The big problem with this system (and the reason it is expensive) is that as the sun moves across the sky, the mirrors have to be turned to keep the sunlight aimed at the absorber. This requires each mirror to be mounted on a pair of axles, and turned at just the right speed (which isn't even constant).

As yet, there doesn't seem to be a 'best' answer for generating solar-electric power, but research is continuing.

Wind power

Windmills can be used to generate electricity. The blades of the windmill (the word 'sails' seems inappropriate for something that looks like an aeroplane propeller) turn a generator, usually through gears to make it turn faster. The big disadvantage of wind power is the difficulty of finding a place where there is a strong, reliable wind all year round, and the fact that generating as much power as even a small conventional power station needs many large windmills.

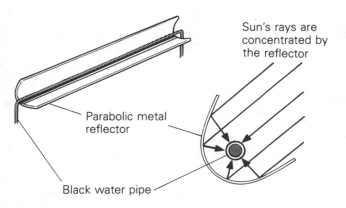

Figure 4.7 A line-axis concentrating collector.

Sun's rays are concentrated by the reflector

Parabolic metal reflector

Black water pipe

Power distribution

It is one thing to generate electric power in a power station; it also has to be distributed to the users. Although the energy requirements of different nations vary a lot, distribution principles are the same the world over.

You will remember that electric power is the product of current and voltage:

$$P = VI$$

To transmit a given amount of power, it is possible to use a low voltage and a large current, or a high voltage and a small current. Even the best conductors of electricity (copper and silver) have a certain amount of resistance. It follows that a long power cable will have a measurable resistance and, according to Ohm's law, there will be a p.d. between its ends if a current is flowing in it.

This results in a heating effect. The amount of power lost in a cable that is several kilometres long can be considerable. The power loss is proportional to the square of the current:

$$P = I^2R$$

where P is in watts, I is in amps and R is in ohms. A few moments' thought (or a few practice calculations) will show that there is considerable advantage in getting the current as low as possible, for minimum wastage of energy. It is also useful to get the resistance of the cable as low as possible, but the only way to do this is to increase the diameter, and thus the weight and cost, of the power line.

It is clear that if we want to minimise wastage, the voltage of the transmission lines should be very high.

High voltages require very effective insulation. If a generator is designed to generate a high voltage (by having many turns on the armature winding), the insulation between the terminals must be excellent. There are limits to what can be done, and

power station generators usually work at below 33 kV. Although this is a high voltage, it is not enough for efficient distribution of power over long distances; in practice, 400 kV is often used as a standard for long-distance lines. The actual voltage varies from country to country, but transmission voltages are always very high.

A transformer (see Chapter 2) is used to increase the voltage to the required level. Transformers work only with alternating current, which is why electricity generated for distribution must be a.c. At such high voltages, the transmission lines would present serious problems of insulation if they were at or near the ground (which is why power transmission lines cannot be buried under the ground). Instead, the lines are supported high in the air, suspended on insulators. The power lines themselves are uninsulated.

At the consumer's end of the distribution lines, the voltage needs to be stepped down to a safe level. This is done in several stages. There are step-down transformers at the end of the long-distance distribution lines that drop the voltage to a few tens of kilovolts. This voltage may sometimes be supplied direct to major industrial users such as steel plants. The lower-voltage distribution lines may also go to a consumer substation where the voltage is further reduced, in most countries to around 400 V, for use by smaller industrial users.

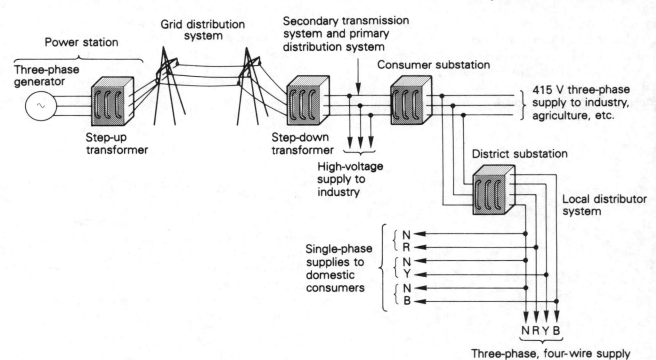

Figure 4.8 A national power-distribution system.

Each district supplied with electricity will have a district substation where the voltage is further dropped, usually to around 240 V or 100 V for domestic users. The three-phase supply is also split into three individual supplies, using a four-wire system. Each phase has a 'live' wire and there is a neutral wire that is common to all three phases. A special transformer at the district substation provides the required conversion. Houses supplied from the district substation are divided between the three phases as equally as possible, to even out the load on the distribution system. Wires connecting the district substation to individual houses may be buried, or may be suspended. A typical distribution system is illustrated in Figure 4.8.

Power and energy

The way in which power is dissipated by resistors and other components has already been covered. You will remember that the SI unit for power is the **watt**, symbol W. Notice that there is no time factor involved. Power is best described as the 'rate of working', and does not depend on time. Nor is the watt an exclusively electrical unit; it is a general unit for power of all sorts. Thus not only electrical power but heat, mechanical and all other kinds of energy are measured in watts.

The older unit of power was the **horsepower**, which you will still find used in a few places. In case you ever need to translate into SI units, one horsepower is equal to 746 watts.

It is useful to have an idea of how much power a watt represents. It is easy to guess what a horsepower might involve, but 1/746th of a horse is hard to visualise! Look at a few common examples, given in Table 4.1. The last item refers to the amount of heat generated if a person is sitting or standing still.

Table 4.1 Some common examples of power consumption.

Light bulb	100 watts
Radio	5 watts
Refrigerator	100 watts
Electric oven	3000 watts
Electric fan	100 watts
Central heating	1800 watts
Microcomputer	25 watts
Calculator	0.001 watts
Person (at rest)	100 watts

If a man is working at a constant rate – let's say he is digging a hole and is using energy at the rate of 700 watts – then we know his rate of work. But to get the hole finished involves working at that rate for a certain amount of time; let's say four hours. The total amount of energy he expends on the job (assuming he doesn't stop or slow down) will be

700 × 4 = 2800 watt-hours, or 2.8 kilowatt-hours

The **kilowatt-hour**, abbreviated to kWh, is a unit of **energy**.

Reference to Table 4.1 will show that 1 kWh represents quite a lot of energy, enough to run a microcomputer for about 400 hours. To measure smaller amounts of energy, such as we may find in electronic equipment, the **watt-second** is more convenient. The watt-second has a special name: it is called the **joule**, abbreviation J, after James Prescott Joule, a British physicist who determined the mechanical equivalent of heat in the 1840s.

If a kilowatt-hour is rather a large unit, then a joule is a rather small unit of energy. As we can see from the table, it would run the microcomputer for only 1/25th of a second!

Measuring the cost of electrical energy

The kWh is used for measuring the amount of electrical energy used by domestic and commercial users of electricity.

In most countries, every house or other user of electricity has the mains electricity supply connected via a **kilowatt-hour meter**. The meter records how much energy is used. Modern meters have a digital readout: that is, the amount of energy used, in kWh, is displayed on a mechanical counter that is very similar to the 'mileometer' (or odometer) that records the distance travelled by a car. Figure 4.9 shows a typical meter of this sort.

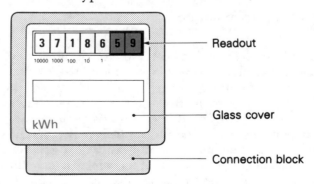

Figure 4.9 A digital electricity meter.

The kWh meter is correctly (but seldom) called an **integrating induction meter**. It is designed to take account of both the supply voltage and the current taken by the consumer, to give a fair basis for charging the customer according to the amount of energy he has used.

When the customer's electricity bill is calculated, a note is made of the reading on the meter. In Figure 4.9, this is 37 186.59 kWh. When it is time for the consumer to pay another bill, the meter is read again. Let us say the reading is 38 159 (it is usual to ignore the decimal). The total energy used in the period between the bills is

38 159 − 37 186 = 973 kWh

This will be multiplied by the charge per kWh (in this context, a kWh is sometimes referred to as a 'unit') to arrive at the cost of the energy that the consumer has used.

Older types of meter use a dial readout instead of a digital readout. A dial readout is simpler to make, but harder to read than a digital readout. The dial readout in Figure 4.10 shows the same reading, 37 186.59 kWh, as the digital readout in Figure 4.9.

Reading a dial meter is easy when you remember that, on each dial, you have to read *the figure that the pointer has just gone past*. Look carefully at Figure 4.10 to see how this is done.

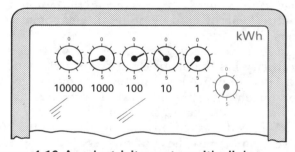

Figure 4.10 An electricity meter with dials.

Overload protection

You have already seen how a fuse or circuit-breaker can be used to protect a circuit in the event of misuse or failure. In mains circuits, such protection is vitally important. Because the voltage is high – usually 240 volts or 100 volts – and the amount of current the power station can deliver is practically unlimited, a fault could quite easily do a great deal of damage. The amount of power that can flow when a mains power supply is short-circuited is enough to cause a fire, or even an explosion.

Fuses are therefore used at several points in any mains supply. Individual appliances are fitted with fuses according to their requirements. In the UK (and in many other countries) every plug connecting an appliance to a wall socket must be fitted with a fuse. In a 240 volt system, small items (such as lamps and radios) will have 1 amp fuses; appliances using more power (such as televisions) will have 3 amp fuses; appliances using a lot of power (such as heaters) will need 13 amp fuses.

Different parts of the supply circuit in the house are equipped with fuses, too. A typical house (220–240 V supply) might have separate fuses for upstairs and downstairs lighting (5 amp), and for power (15 amp). If a cooker is fitted it should have its own special circuit with a 30 amp fuse.

Finally, the point where the supply from the power station's cable enters the house will be protected by a single heavy fuse, typically 60 to 80 amps, as a last line of protection for the supply.

This sort of system gives three levels of protection. A faulty appliance will generally blow its own internal fuse if it has one, or the fuse in the plug. If the wiring is faulty, or if there is no plug fuse, the circuit fuse will blow. And if the circuit fuse fails for some reason, the supply fuse will blow.

Circuit-breakers (see Chapter 2, Figure 2.15) are used in the most modern installations in place of the circuit fuses. They have the advantage that they are more reliable, and can be reset by simply pressing a button.

A useful form of protection is a device called an earth-leakage circuit-breaker, which compares the current flowing in the live and neutral wires, disconnecting the power if there is more than a very small difference, about 30 mA.

D.C. electric motors

A d.c. electric motor is very similar in construction to a d.c. electric generator. In fact, most types of electric motor will generate a voltage if you spin the shaft.

The simplest sort of d.c. motor is the **permanent-magnet motor**. Permanent magnet motors are used in all sorts of low- and medium-power applications. Electrical appliances such as tape recorders and hairdriers use permanent-magnet motors. A typical permanent-magnet motor is illustrated in Figure 4.11.

When a voltage is applied to the terminals, the

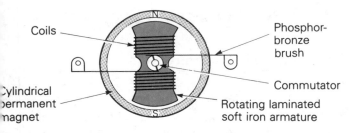

Figure 4.11 A permanent-magnet d.c. motor.

armature coil produces a magnetic field in the armature. This field causes the armature to rotate so that it tries to line up with the north–south axis of the ring-shaped permanent magnet surrounding the motor. Just as the armature reaches alignment, the commutator switches the polarity of the connection to the coil. Momentum takes the armature past the alignment point, but now the armature, with its magnetic field reversed, moves towards alignment on a south–north axis. As it reaches this point, the commutator again switches the connections. Thus the armature continues to rotate for as long as current is supplied to the motor.

The simplified motor illustrated has one winding, like the simplified generator in Figure 4.1. Because the armature has two poles (which can be north or south, according to the position of the armature and polarity of the applied voltage) it is called a **two-pole motor**. Figure 4.12 shows an end view of a two-pole motor and a four-pole motor. All else being equal, the four-pole motor will be more powerful.

We saw that the magnetic field in which the armature of a generator rotates can be produced by a permanent magnet or by an electromagnet. The same is true of an electric motor, and for larger sizes, electric motors use electromagnets to produce the magnetic field in which the armature spins. The coil that produces the electromagnetic field, called the **field winding**, can be connected either in parallel with the armature winding, or in series with it. Both methods are used, and both

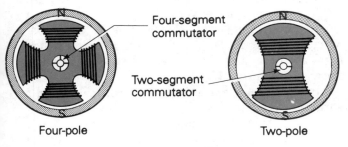

Figure 4.12 Two-pole and four-pole motors.

have advantages and disadvantages. It is useful to compare the three main types of d.c. motor.

Permanent-magnet motors

Permanent-magnet motors are widely used where small, cheap, low- to medium-power motors are required. The working voltage is usually low, from 1 V to 50 V. Operating current can be from a few milliamps to a few tens of amps according to size. It is possible to control the output power, and thus the speed, of a permanent-magnet motor by means of a resistor in series with its power supply. Accurate control is possible with specialised electronic circuits. The permanent-magnet motor, unlike other types, can be reversed simply by swapping over the power supply polarity. Permanent-magnet motors cannot be used with a.c.

Series-wound motors

The field winding is connected so that it is in series with the armature winding, and all the current used by the motor flows through both windings. The field winding is therefore made with a low resistance (as few turns of thick wire as possible). This kind of motor can be made in almost any size. In the larger sizes, speed control is expensive, requiring electronic power control devices that will withstand high voltages and large currents. Typical applications range from medium power (electric mixers, car starter motors) to high power (electric locomotives).

Because the magnetic field of the armature and the magnetic field of the stator are both produced by electromagnetism, the motor will run in exactly the same way regardless of the polarity of the supply voltage. One might therefore expect such a motor to work with an alternating current supply in which the polarity is continually changing and this, in practice, is generally true. Most series-wound motors will work with d.c. or a.c. supplies.

It follows that changing the supply connections will not reverse this kind of motor. A series-wound motor can be reversed only by changing over the connections between the armature and field windings.

Shunt-wound motors

In a shunt-wound motor the field winding is connected directly across the supply. The field winding is usually made with a high resistance (lots of turns of fine wire) so as to generate the strongest possible magnetic field while using the least amount of current. Because the current is additional to the armature current, the shunt-wound motor is less efficient than the series-wound motor in the amount of power it produces for a given current. But because the field coils draw a relatively small current it is fairly easy to control the speed of a shunt-wound motor. A simple variable resistor in series with the field winding will provide a useful speed range.

A shunt-wound motor can usually be run on a.c. or d.c. It can be reversed by changing the polarity of the field winding (but not that of the armature winding).

Self-inductance and problems of commutator design

Small permanent magnet electric motors have very simple commutators, consisting of copper and brass segments, with flat bronze springs – the brushes – carrying the current to them. On larger motors, commutator design is more of a problem.

We saw in Chapter 2 that when a current flowing through a coil is interrupted, the energy that has gone into the creation of the magnetic field produces a pulse of e.m.f., usually at high voltage. In a motor or generator, self-inductance causes pulses of high voltage to appear as the commutator switches the current from coil to coil in the armature (or reverses the polarity if it is a two-pole motor). The high voltage causes electric sparks. Sparks are bad for two main reasons:

1. they burn away the commutator and brush;
2. sparks are powerful emitters of radio interference.

Various methods are used to reduce, or **suppress**, sparking at the commutator of a motor. Brushes are often made of carbon, because it is conductive, compliant and, although it can be burnt by sparks, it will still conduct. Large motors have carbon brushes that are designed with erosion caused by sparks in mind, having springs that push a long carbon rod (the brush) towards the commutator to compensate for wear.

In multiple-pole motors, brushes may be designed to cover more than one segment of the commutator at a time.

A.C. electric motors

Three-phase induction motors

Although d.c. motors other than permanent magnet motors can usually be used with a.c., it is possible to design motors that work more efficiently with a.c. In particular, it is possible to make them without commutators, thus removing one of the most difficult design problems and the only source of radio interference.

You will remember that when we looked at generators we discovered that if we moved a conducting wire through a magnetic field, a current was induced in the wire. A current-carrying wire, placed in a magnetic field, experiences a force that is at right-angles to the magnetic field. So if we move a magnetic field across a wire (or across a coil, to increase the effect), two things happen:

1. a current is induced in the wire by the moving magnetic field;
2. the current flowing in the wire tends to move the wire at right angles to the field lines.

The result is this: the wire tends to be pulled along in the direction of the moving field.

This is a very useful principle, because it means that if we can contrive a magnetic field that rotates, we can make an electric motor that requires no connections to be made to the armature winding. The moving magnetic field induces the armature current, without the need for a commutator or slip rings.

If we are using a three-phase mains supply, a rotating field is easy to arrange. The motor is made with three field coils, each field coil connected to one phase of the supply. The magnetic field produced by the three phases (each, you will recall, 120° apart) appears to rotate as the peaks and troughs of the three sine waves appear on each coil in turn. The coils are in effect reproducing the magnetic field 'seen' by the rotor of the generator.

The armature needs very thick wires to carry a heavy current at low voltage; this is the most efficient form. In fact, several single-turn 'windings' are best, and the design most often used is shown in Figure 4.13.

For some reason, this design of rotor is called a **squirrel cage** (although I have never seen a squirrel kept in a cage that looks anything like it). The rotor bars (Figure 4.13(a)) are made of thick copper and are solidly welded to copper end rings that

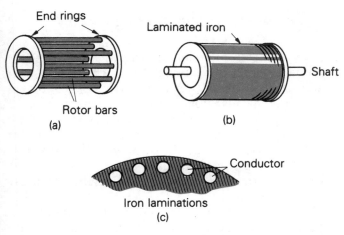

Figure 4.13 Construction of the rotor of a squirrel-cage induction motor.

complete the electrical circuit. Practical motors use a large number of thin iron sheets, or laminations, threaded over the bars, to improve the magnetic efficiency of the rotor by concentrating the magnetic field (Figure 4.13(b) and 4.13(c)). We now have all the parts for an a.c. motor with no commutator (Figure 4.14). The rotating field induces a very large electric current to flow in the rotor bars and end rings of the rotor. This current produces its own magnetic field, which drags the rotor round in the same direction as that in which the magnetic field is rotating.

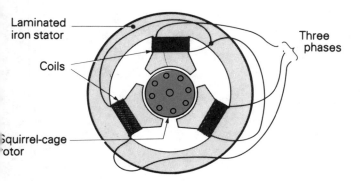

Figure 4.14 An a.c. motor with no commutator.

In countries where the mains supply is at 50 Hz, the magnetic field rotates at 3000 revolutions per minute (or 3600 rev/min in countries that use 60 Hz supplies). The rotor can never actually reach this speed. The reason is simple. At 3000 rev/min the rotor would be stationary with respect to the rotating field, and no current would be induced in the rotor. The amount by which the rotor fails

to 'keep up' with the rotating field is called the **fractional slip** of the motor, and would be expected to be between 2 and 10% for most motors.

Single-phase induction motors

Although three-phase induction motors are widely used in industry, the single-phase supply provided for domestic installations means that they cannot be used in the home.

It is difficult to produce a rotating field with a single-phase supply but it is easy to produce a field that increases and decreases, and thus use the principle of the transformer (Chapter 2) to induce a current in the rotor. Unfortunately, there is no reason why a motor built on this principle should ever start running. Once the rotor is spinning, its own field – which is of course rotating – keeps the motor going; but something extra is needed to get it moving.

In medium and large motors, a second winding is used to start the motor, connected through a capacitor. The time taken by the capacitor to charge and discharge causes the field produced by what is known as the **starter winding** to be phase-shifted in relation to the main winding, and this provides a rotating field to get the motor moving. The starter winding should not remain connected, so motors with this kind of starter have a centrifugal switch, that disconnects the starter winding when the motor approaches its running speed.

Smaller motors use a **shaded-pole** rotor, the details of which are beyond the scope of this discussion. However, the shaded-pole rotor uses a copper short-circuiting ring in the rotor to twist the magnetic field, which starts the rotor moving. Shaded-pole motors have less power at start-up than motors with a separate starter winding, and usually have to be disconnected from their mechanical load for starting.

Other types of motor

There are many different varieties of motor, for hundreds of different applications. Those mentioned above are the main types. Others that you should know about, but without going into details of their operation, are: **synchronous motors**, in which the rotors follow exactly the mains frequency (used in mains-powered clocks, etc.); **stepping motors**, which allow exact control of the rotation of the shaft (used in computer applications); and **linear motors**, which are in effect induction motors 'unrolled' so that a linear motion along a special track is obtained instead of a rotary motion.

■ CHECK YOUR UNDERSTANDING

● 'Generators' in power stations are **alternators** that produce alternating current at a frequency of either 50 Hz or 60 Hz.

● The output of an alternator is a **sine wave**.

● Power stations are often powered by **steam**. Steam can be generated by coal, oil or natural gas (fossil fuels), by atomic fission, or by heat from the sun.

● Generators may also be powered by falling water (hydroelectricity), or by windmills.

● Electricity is distributed along power lines at very high voltages so as to minimise the amount of power lost because of the electrical resistance of the lines.

● Electric **motors**, like generators, make use of the magnetic effect of an electric current. They convert electrical energy into mechanical (almost always rotational) energy.

1 Draw a sketch of a simple d.c. generator.
2 Draw a sketch of a simple alternator.
3 Write a short essay (not more than a page) comparing the advantages and disadvantages of atomic power, oil and hydroelectricity as sources of energy for making electricity.
4 The horsepower is a unit of power. The watt is another unit of power. What is
   i) the relationship between the horsepower and the watt, and
   ii) the unit of energy derived from the watt?
5 Permanent-magnet motors are very common in all sorts of applications that require small motors. What is the main advantage over a shunt-wound motor?
6 Why is electricity distributed via overhead high-voltage lines, when it would be less unsightly to have cables buried under the ground?

Electrical installations

Introduction

In any domestic or industrial installation, the first consideration is safety. This means that the wiring must be correctly designed (for example, heavy enough cables for the load being carried) and correctly installed.

Most countries have detailed regulations for the installation of electrical wiring, and it is essential that these rules are followed. Since this book is intended for readers in more than one country, I shall deal only with the principles of installations here, and not with details of regulations.

> Any reader of this section should get a copy of the relevant wiring regulations, and become familiar with them.

Domestic power supplies

Whether the power to a house is supplied through underground or aerial cables, there will always be a main fuse, switch, and distribution fuseboard at or near the point where the cable enters the building. Figure 5.1 shows a typical intake for a home.

The features to note are:

1. The heavy fuse (in 220 V systems it will be 60–80 A) in the live connection to the input. This is the 'last link' in the chain of protective devices for the supply.
2. The electricity meter, to measure the amount of energy used, in kWh.

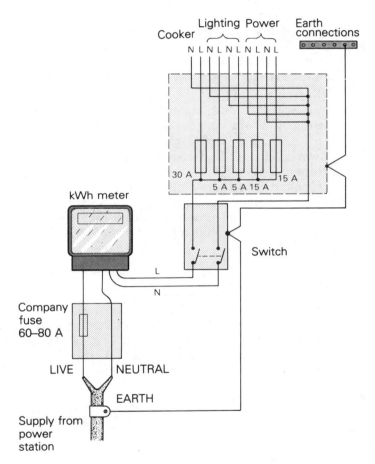

Figure 5.1 A typical intake for a domestic power supply.

3. The main switch: this may be fitted with its own fuse. Note that this is always a double-pole switch and disconnects both the mains and neutral lines of the supply.
4. The distribution fuseboard. This has a number of fuses (rated at up to 30 A in a 220 V system). It should be wired in a logical way, dividing the house into separate power circuits: upstairs

lighting, downstairs lighting, upstairs power, downstairs power, cooker.

In the more modern fuseboards, circuit-breakers are used instead of fuses, as they are easier to reset and can be adjusted so that they break the circuit at a more accurately controlled point.

5. The earthing lead. This heavy cable (in the UK, a minimum of a 6 mm² conductor) provides an earth return path for fault currents in the circuit.

In a house, there are three main classes of wiring: lighting circuits, power circuits and cooker circuits.

Electric lights take relatively little current; a 100 W lamp on a 220 V supply consumes only around 0.5 A. **Lighting circuits** are designed to have a maximum loading of 5 A in most countries using a 220–240 V supply, or 10 A in countries using a 110–120 V supply.

Power circuits for the connection of electrical appliances including heaters require a large current capability. Typical in countries with 220 V supplies is a 13 A or 15 A maximum. This is enough for a powerful electric heater, and almost any appliance except a cooker.

Cooker circuits are designed specifically to serve an electric cooker and nothing else. Because cookers contain many heating elements they use a lot of power, and a circuit capability of 30 A at 220 V is needed.

Lighting circuits

The requirement of a lighting circuit is always much the same: a switch, plus one or more lamps, often at a distance from the switch. A typical lighting circuit for a room is shown in Figure 5.2.

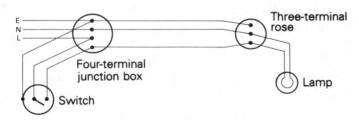

Figure 5.2 A typical room lighting circuit.

A four-terminal **junction box** is needed at the point where the switch is connected to the power line. Junction boxes for domestic use are circular, made of plastic, and fitted with a number of screw terminals. They are available in various sizes, to

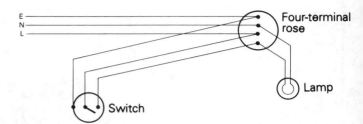

Figure 5.3 A room lighting circuit with a four-terminal rose.

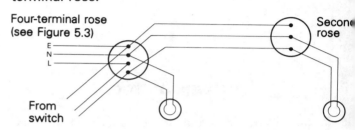

Figure 5.4 Parallel wiring for two room lamps.

Figure 5.5 Two-way switches.

take different weights of cable, for lighting or power applications.

Since a light fitting on the ceiling of a room will itself need a junction box, known as a **ceiling rose**, it is more economic to use a four-terminal rose which is used for the connections to the switch as well as to the lamp. This is shown in Figure 5.3. If two lamps are needed, they are simply wired in parallel, as in Figure 5.4. Sometimes two switches are needed, so that light can be controlled from two positions. In this case two-way switches are used, connected as shown in Figure 5.5.

If you want to be able to adjust the brightness of lights, a lighting controller can be used in place of the switch, or one of the switches in a two-way system. The connections are exactly the same, but the controller has a knob allowing the brightness of the lights to be controlled from 'full' down to a very low level. It is not possible to use a lighting controller at both positions in a two-way system.

In countries that have a 220 V supply, all lighting circuits are wired with 5 A cable.

Power circuits

Figure 5.6(a) shows two 13 A, 240 V power plugs, wired in parallel and connected to a supply. If an electric kettle is plugged into each, and each kettle takes 10 A from the supply, the wire at point A is carrying 20 A, as the power to both sockets is flowing through it.

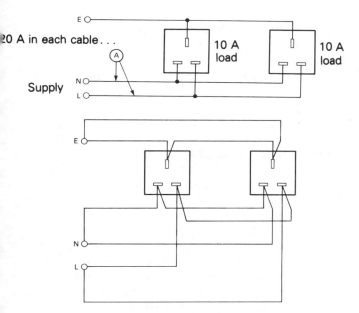

Figure 5.6 (a) Parallel connection of two sockets; (b) two sockets on a ring main.

A cable that carries 30 A (two lots of 13 A, plus a small safety margin) is large, about as thick as a man's index finger, and expensive. In modern installations a system is used that enables 15 A cable to be used and which also has other advantages. This is the **ring main**.

A ring main is illustrated in Figure 5.6(b). The name is descriptive of the way it works. Each of the three connections to the sockets (live, neutral and earth) starts at the distribution fuseboard, goes round each socket and then back to the fuseboard. Current is conducted to the sockets through both parts of the ring, so each socket is effectively connected to the supply through two cables, connected in parallel. Ring mains are a requirement of the wiring regulations in most countries. Where there is a 220 V supply, ring mains are wired with 15 A cable.

Cooker points

A cooker point should be connected to the distribution fuseboard by a single length of cable with no joins in it. Heavy-duty cable of at least 30 A capacity must be used for 220 V systems. The cooker point itself is simply a heavy double-pole switch, for isolating the cooker from the supply. A short length of heavy cable connects the cooker to the point.

> You should not connect anything other than a cooker to the cooker point. Where there is more than one cooker, each should have a separate cable run and a separate fuse at the distribution fuseboard.

Wiring practice

PVC sheathed cable

Domestic wiring, and a lot of office and industrial wiring, is done with **PVC sheathed cable**. PVC is the abbreviation for a plastic called **polyvinyl chloride**, which is used for insulating and protecting wires. The most common type of PVC sheathed cable is illustrated in Figure 5.7. The cable in the illustration is usually referred to as 'two core and earth' cable because the centre conductor has no insulator other than the sheath. National wiring regulations often call for this type of cable to be used in house wiring. The two insulated conductors are insulated with PVC sleeves, and are colour-coded red and black.

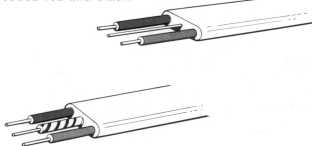

Figure 5.7 'Two-core and earth' PVC sheathed cable.

> ▲ It is important to treat PVC cable properly when carrying out installations. When stripping the insulation off the cable, it is vital not to cut through the insulating sleeves by mistake and equally vital not to nick the copper conductors themselves.

'Two core and earth' cable is best stripped as follows. With a sharp knife, make a cut along the centre of the side of the sheath, using the unsleeved earth conductor as a guide. Pull the sheath back to the end of the cut and trim off the unwanted end. Now, using light pressure only, make an incision around the sleeve where you want to remove it, and twist and pull off the end of the sleeve with your fingers. Don't be tempted to cut the sleeve using more pressure, or you will nick the conductor, which will reduce its current-carrying capacity at that point and could lead to overheating or failure. A variety of wire-stripping tools are available; any of them makes the job of stripping the sleeve easier. Always take a careful look at the wires when you have stripped them and, if you have damaged the conductor, do the job again.

Most modern cables for house wiring have a single copper conductor, although stranded cables (which are more flexible) are often found. The solid-core cables should not be bent too sharply or the insulation or conductors could be damaged. A good guide is that 5 A cable should not be bent round a sharper radius than 25 mm, and 15 A cable should not be bent round a radius sharper than 60 mm.

Take care when connecting the cables to a junction box. All screw terminals should be securely tightened, and the sheath should extend inside the box. Figure 5.8 shows the right way and the wrong way to do it. Junction boxes should not be left 'hanging' by the wires, but should be screwed to a suitable surface: a wall, joist or batten.

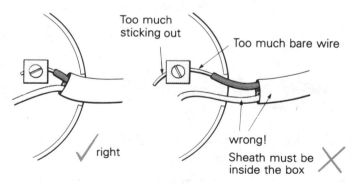

Figure 5.8 The right and the wrong way to connect cables to a junction box.

Switches are usually flush-mounted so that the switch plate is level with the wall. Steel switch boxes are available for fitting into the wall before the wall is plastered. The switch box is usually nailed in place, then finished with plaster after the electrician has completed the wiring. Special boxes can be obtained for use with plasterboard partition walls; these boxes are self-securing, as it is not possible to nail or screw them into position. Surface-mounting boxes are also made, usually of plastic, to stand proud of the surface to which they are fixed. You would normally use a surface-mounting box where flush mounting is not possible.

Conduit systems

When wiring needs to be protected from the environment, either because of weather, safety considerations or to protect it from mechanical damage, a **conduit** system is used. Conduits are essentially pipes, down which wiring can be run.

Conduits can be made from steel or plastic. Plastic conduit systems provide less mechanical protection than steel conduits (they can be broken, and even attacked by gnawing animals and some insects), but provide better protection from the weather because they do not rust.

Various commercially made systems can be obtained. The steel systems usually consist of cast or fabricated joint boxes, which have one or more threaded spout entries into which threaded conduit pipes can be screwed. It is usually up to the electrician to cut suitable threads on the end of the conduit when he has cut it to length. Steel conduit systems must always be earthed, to provide a return path if one of the wires inside should become shorted to the conduit.

Essential tools for steel conduit work are a solid workstand (a 'tripod vice'), a pipe-bending machine or wooden pipe-bender so that the conduit can be bent without flattening it out, thread-cutting dies, a stilson wrench for gripping conduits, and a reamer or file for smoothing the sharp edges of the conduit where it has been cut.

Plastic conduit systems are much the same as metal ones to use, except for the following points.

1. Plastic conduit cannot be bent very much, so the manufacturers make a variety of curves.
2. The constituent parts of a plastic conduit system are usually held together by adhesive.
3. As PVC is an insulator, plastic conduit does not, of course, need to be earthed.

A conduit must never be completely filled by the cables it contains. Regulations usually require that the conduit has a **space factor** of 40 per cent. This means that only 40 per cent of the cross-sectional area of the conduit may be filled by the cables.

A rather specialised form of conduit, called **trunking**, is sometimes used in factories and offices. A trunking system is like a conduit system, but the conduits are large, usually in sections, and fitted with removable lids. This enables wiring to be changed or added relatively easily. Trunking usually has a higher space factor than conduit, perhaps 45 or 50 per cent depending on the national regulations. Various systems are available commercially, for a range of applications.

The wiring of terminals and plugs

Most fittings, from plugs to junction boxes, have a mechanical connector consisting of a screw terminal like the one illustrated in Figure 5.9. When gripping a cable in such a connector, the following rules must be obeyed.

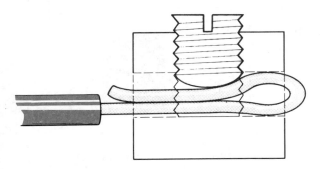

Figure 5.9 A screw terminal, common to many electrical fittings.

1. The conducting core must be undamaged; in particular, it must not have any nicks or cracks. If a multistranded conductor is used, all the strands must be twisted together and clamped under the screw. If any strands are broken off, the termination must be remade. If the core is a loose fit under the screw, it can be folded double.
2. The sleeving must be taken as close as possible to the terminal, but must never be clamped under the terminal.
3. If there is a sheath on the cable (as there would be if making a connection to a junction box) the sheath must be taken inside the box.
4. The terminal must be large enough to accommodate all the conductors easily.
5. The screws must be securely tightened.

Lighting

Incandescent lamps

Lighting was one of the very first uses for electric power. An American, Thomas Edison, developed a workable electric lamp and set about marketing both the lamp and electricity as a form of power. Although Edison's first system worked on d.c. (and he had serious problems with distribution as a result) his lamps worked well and were the technological marvel of the age.

A modern incandescent lamp is not very different from Edison's lamp of a century ago. He would have recognised it immediately. It consists of a filament made of **tungsten** (Edison used carbon) inside a glass bulb containing an inert gas (usually argon) under low pressure (Edison used a vacuum in his bulbs). When a current is passed through the filament, the resistance of the filament causes power to be dissipated in the form of heat: so much heat that the filament glows white hot and gives off light. In air, the filament would burn up in an instant, but the inert gas will not support combustion and the filament remains intact.

Tungsten lamps are made in a variety of shapes and sizes; mains-powered lamps are in the range of 15–1500 watts power consumption. (Torch bulbs and car headlight bulbs are also tungsten lamps, but are designed for lower voltages and power consumptions.) The modern mains-powered tungsten lamp (the common light-bulb) will burn for about 1000 hours before the filament finally breaks because of evaporation of the tungsten.

Tungsten lamps designed for mains operation have built-in fuses, so that when the lamp fails it does so 'quietly', without blowing the circuit fuse, which would put out all other lamps on the same circuit. Clear bulbs and frosted (pearl) or opal (white) versions are commonly available.

There are two types of lamp base: bayonet cap (BC for short) and Edison screw (ES). There are no particular advantages in either type. BC is common in the UK, ES is popular in Europe and America. Some countries use a mixture of both. Figure 5.10 shows the parts of a typical incandescent lamp.

The fitting into which the bulb is inserted is called the **lampholder** and the light fitting itself is correctly called a **luminaire**. Luminaires can be designed to disperse the light evenly or concentrate it on a small area. No luminaire can increase the

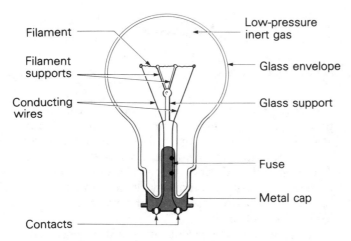

Figure 5.10 An incandescent lamp.

amount of light coming from the bulb, although some use reflectors to concentrate it on, for example, a desk or machine.

Lamps can be obtained that have built-in reflectors to direct all the light forward. These are called **spotlight bulbs**; they are efficient but expensive compared with normal light-bulbs.

Where high brightness or extra power is needed, a modified form of incandescent lamp, the **tungsten–halogen lamp**, is suitable. Tungsten–halogen lamps use a tungsten filament just like normal lamps but they have a much smaller envelope so that the envelope sides are close to the filament. Because a glass envelope would melt, the envelope is made of quartz. The lamp is filled, not with an inert gas, but with a halogen gas, usually bromine. The filament is run at a very high temperature and it evaporates at a rate that would in a normal lamp cause failure within a few seconds. In the tungsten–halogen lamp, however, the evaporated tungsten combines with the halogen to form a compound that has the property of depositing tungsten on a hot surface. This replaces the tungsten lost by evaporation so the filament is continuously renewed.

Tungsten–halogen lamps run hot and must not be overcooled or the renewal cycle breaks down and the filament burns through. Also, the quartz envelope must be completely clean; even the grease left by a fingerprint will burn through the lamp eventually.

Fluorescent lamps

Fluorescent lamps are very popular in shops, offices and factories. They are efficient in that they give a lot of light for the amount of power consumed. The light is diffuse and well suited to general illumination of a large area. Because of their higher efficiency, they give out less heat than the equivalent incandescent lamp.

Fluorescent lamps have some disadvantages. They do not light up immediately. There is a delay between switching on and lighting up that can last from one second to four or five seconds, depending on the type. They cannot be used with the usual lighting controllers, so they can be used only at full brightness. The fittings are more expensive than those needed for incandescent lamps. The main parts of a fluorescent lamp are shown in Figure 5.11.

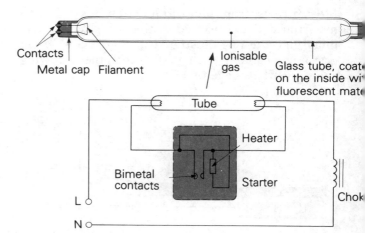

Figure 5.11 A typical fluorescent lamp.

The tube itself is made of glass, and is coated on the inside with a fluorescent material (a **phosphor**) that glows brightly when struck by electrons. At the ends of the tube are filaments that are specially coated to emit electrons when they are heated. The tube is filled with an ionisable gas, such as mercury vapour.

The **starter** consists of a switch operated by a bimetallic strip and a small heater. The switch is initially open, but when current flows through the switch the heater closes the switch by making the bimetal bend. This also short-circuits the heater, so after a moment the switch opens again, and the cycle repeats. There are other types of starter switch in use, but this is the simplest and most popular.

Moments after a fluorescent lamp is first switched on, the starter switch closes. Current flows through the circuit and heats up the filaments at the ends of the tube so that they emit electrons.

After a second or so, the switch opens (see above). The current stops flowing through the inductor (it is usually called a **choke**) and a high-voltage pulse (caused by self-induction) appears between the two ends of the fluorescent tube.

This ionises the gas inside the tube, which becomes able to conduct current. Current now flows through the inductor, which limits the current flow, and through the length of the tube. The starter is short-circuited by the ionised gas in the tube, which has a low resistance, so the starter switch does not operate again. Note that the filaments in the fluorescent tube are not allowed to cool, as one terminal of each filament is disconnected.

If the gases in the tube fail to conduct (for example, if the tube is very cold) the starter will operate repeatedly until the tube 'strikes'. The electronic activity in the tube causes the fluorescent material lining the tube to glow brightly, emitting light.

Switches

There is little to be said about switches, except to note that you should always use a switch designed for the purpose. Mains switches are made in various sizes and power ratings, for lighting, heating and power control.

It is worth mentioning that modern switches are very different from those produced 20 years ago. It used to be thought that it was best to break the current by moving the switch contacts as far apart as possible in as short a time as possible. Modern switches are of the slow-break microgap design, in which the contacts move apart rather slowly, and the total distance they move is less than a millimetre. They are far more reliable than the old sort.

> Single-pole switches are always connected in the live line. Double-pole switches are connected in the live and neutral lines. The earth line is never switched.

Wiring regulations

As I said at the beginning of this chapter, it is essential to refer to national wiring regulations before tackling electrical installations. In general, good practice is largely a matter of common sense. If you try to bear in mind the *reasons* for what you are doing, then you will remember all the better. For example, why is it important to ensure that the sheath of a wire rather than the cores is under the clamp in a plug? The answer is, of course, that the sheath takes the mechanical strain, and also provides a second layer of insulation. If you put a plug on a lead in such a way that you can touch the insulating sleeves around the cores, you could be risking your life by betting that a millimetre of PVC will not break, even after months of wear.

You should be aware that fire is probably more of a danger than electrocution. If you leave a screw terminal loose, a wire could slip out and short-circuit, causing a fire. A cable that is carrying too large a current will overheat and could start a fire.

Finally, you must take care of yourself. Never work on any circuits unless you are completely sure that they are properly disconnected. Don't trust your life to someone else's work by relying on a single-pole switch to disconnect a circuit. They may have connected it in the neutral line instead of the live line, leaving the circuit in a lethal state even when switched off. Always turn off the power at the main switch.

■ CHECK YOUR UNDERSTANDING

● **Electrical installations** are subject to national regulations, which must be followed if the installation is to be safe and legal.
● Domestic power supplies are always fitted with a large fuse (the 'company fuse') which is the final link in a number of safety fuses.
● Domestic supplies can be isolated from the electricity supply by the **main switch**, a double-pole switch that disconnects the live and neutral wires.
● The **distribution board** has a number of fuses or **circuit-breakers** that can interrupt supplies to different parts of the building if there is a fault.
● Domestic power is divided up into **lighting**, **power** and (usually) a **cooker** supply. Each has

separate fuses. (As an exercise, find out the fuse ratings in your own building.)

● **Conduit** systems are used where wiring needs extra protection. Conduits are made of either metal or plastic.

● **Incandescent lamps** (the common light-bulb) work by heating a tungsten filament in an inert gas. The filament gets white-hot and glows brightly.

● **Fluorescent lamps** light up because electricity flowing through an ionised gas inside them makes the phosphor coating on the inside of the tube glow. They give more light than an incandescent lamp with the same power consumption.

REVISION EXERCISES AND QUESTIONS

1 What are the main fuses in a domestic power installation likely to be?

2 Most domestic room-lighting circuits are now wired with four-terminal roses. Sketch a simple one-light installation using this method.

3 What are the advantages of fluorescent lights compared with incandescent lamps? What are the disadvantages?

4 Give *two* examples of colour codes for mains wiring.

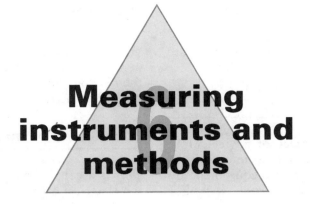

Measuring instruments and methods

Introduction

It is, of course, impossible to see an electric current or voltage. We can detect electrical quantities only by their effects: heating, lighting, chemical, magnetic and mechanical. A range of instruments are used by the electrical and electronics engineer to tell him what is going on in a circuit. In this chapter we shall consider a variety of different measuring instruments and the way they are used.

When using any instrument, there are aspects to be considered under two headings:

1. Safety
   A) your own;
   B) that of the circuit being tested;
   C) that of the measuring instrument.
2. Accuracy
   A) inherent accuracy of the instrument;
   B) accuracy of the way you take the measurement;
   C) the degree of accuracy you actually need.

Safety

Personal safety

Your own safety is obviously very important. When working on any mains electric wiring or appliance it is vital that you disconnect the supply before you start. This may mean pulling out a plug, turning off the main switch, or removing a fuse. If you are working on house wiring it is never safe to rely on wall or power-point switches; they may fail, or they may be incorrectly wired into the neutral line. **Never take chances.** It is hardly ever

necessary to make measurements on mains wiring with the supply connected.

Most – but by no means all – modern electronic circuits work on relatively low voltages and are inherently safe. Some common appliances use dangerously high voltages: televisions are a good example of this. Where an electronic circuit is supplied with a low voltage by a built-in mains transformer (such as most radios, hi-fi systems, and mains-powered tape recorders) then you should *not* rely on the transformer for safety: it could have faulty insulation.

> ⚠ Always use an isolation transformer when working on mains-powered equipment (see Chapter 1). This gives the best measure of protection.

Safety of the item under test

It is unlikely that you will do much damage to mains wiring, switches, or fittings unless you are unusually clumsy. However, some electric motors – such as those in washing-machine pumps and food mixers – are not designed to be run for more than a few minutes. If during the course of testing you were to run such a motor for a long time, parts of it might overheat and even melt.

Electronic equipment is more delicate. It is easy to damage most electronic circuits by connecting them to a supply voltage that is too high, or applied with the wrong polarity (positive and negative swapped over). Some kinds of **integrated circuit** (such as those used in computers) can be wrecked just by touching them, so don't investigate any electronic circuit unless you know what you are dealing with.

Safety of the instrument

'Don't drop it' is the first piece of advice. More test meters are broken this way than in any other. It is also possible to damage test equipment by carelessly connecting it in the wrong way. For example, a meter designed to measure electric current, if connected directly across a battery, will probably not survive.

Accuracy

Accuracy of the instrument

No item of test equipment is 100 per cent accurate. A test meter may well have a basic accuracy of ±10 per cent. Meters that have a basic accuracy as good as ±2 per cent are likely to be very expensive. Fortunately, practical electrical and electronic circuits seldom call for very accurate measurements.

Accuracy of your measurement

It is a basic law of physics that you can't measure something without affecting it in some way. It is worth remembering this. For example, if you put a meter designed to measure current into a circuit, then the presence of the meter itself will affect the current flowing. Whether it affects the measurement significantly depends on the design of the meter and on what you are trying to measure with it.

Accuracy that you need

A test meter may show a reading of (say) 0.2 V. If the meter has a basic accuracy of ±10 per cent, and the sensitivity of the meter (see below) is fairly low, then you might expect the true value of the p.d. to be between about 0.15 V and 0.22 V. Bearing this in mind, you have to ask yourself, 'Does this matter?'. If you were expecting the voltage to be somewhere about 0.1–0.25 V before you measured it, then the answer is that it would not. If, however, you were measuring the voltage at a critical point in the circuit and were expecting 0.15 ±0.01 V, then it does matter, and the meter you are using is not accurate enough to give a conclusive answer.

Quantities to be measured

This is a convenient point at which to summarise the electrical quantities that we have encountered:

Potential difference	volt	V
Current	ampere	A
Resistance	ohm	Ω
Capacitance	farad	F
Power	watt	W
Energy*	joule	J
Inductance	henry	H
Frequency	hertz	Hz

*In installations and power, the more common measure of energy is the kilowatt-hour (kWh).

Meters

Today, both **analogue** and **digital** meters are available, and both are in common use. The earliest measuring instruments were analogue in nature, and we shall begin by looking at this type.

An analogue meter uses a pointer, needle, or other indicator to point to a scale that is calibrated in volts, amps etc. The dictionary definition of the word 'analogue' is '*Analogous or parallel word or thing . . .*'; the movement of the pointer of an analogue meter moves in sympathy with the quantity being measured. The larger the quantity, the further along the scale the pointer moves.

Moving-coil meters

Most analogue meters use **movements** (the movement is the mechanism that moves the pointer) that are based on magnetism. Figure 6.1 shows the most common design, the **moving-coil** movement.

The principle is simple. The coil and meter pointer are mounted on bearings (often jewelled, like a watch) so that the coil can rotate. Two small **hairsprings**, like the balance of a clockwork watch, resist the rotation and normally hold the coil in a fixed position, with the pointer at one end of the scale. The coil is located in a strong and constant

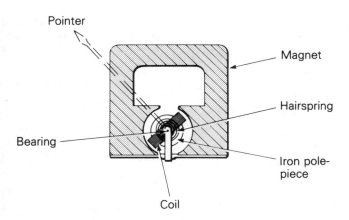

Figure 6.1 A moving-coil meter.

magnetic field, that of a permanent magnet.

When a current is passed through the coil, a magnetic field is created by electromagnetism. The orientation of the field is such that it rotates the coil against the springs. The amount that it moves depends upon the strength of the electromagnetic field, which in turn depends upon the amount of current flowing through the coil. It is convenient to use the hairsprings themselves to carry current to the coil.

Moving-coil meters are made in many shapes and sizes, according to the application. Electrically, they are specified according to three main parameters: coil resistance, sensitivity and accuracy.

The coil **resistance** is just what it sounds like: the electrical resistance of the moving coil, in ohms. This can be any value from just a few ohms to several kilohms.

The **sensitivity** of a meter is often quoted as the full-scale deflection, or FSD for short. It is measured in amps (or more often milliamps or microamps), and is the amount of current that has to flow through the meter coil to make the pointer move to the far end of its scale.

Another useful measure of sensitivity, which includes the resistance of the coil (FSD doesn't) is given by

$$S = \frac{R_\mathrm{M}}{V_\mathrm{FSD}}$$

Here R_M is the resistance of the meter movement coil, and V_FSD is the voltage required to produce full scale deflection. The sensitivity, S, is given in **ohms per volt**. This parameter may also be quoted by manufacturers.

The **accuracy** of a meter is given as a percentage

tolerance, in just the same way that a component such as a resistor has a tolerance value. A meter with a sensitivity of 1 mA and an accuracy of ±20 per cent is a meter that will give a full-scale deflection of its pointer for a current ranging from about 800 μA to 1.2 mA. This does not mean that the meter will vary by that amount from time to time, just that the meter may depart from its specification by ±20 per cent. The repeatability (a fourth parameter!) of measurements will be much better than this, and is related to the mechanical construction of the meter: the quality of the bearings, for example.

Moving-coil meters are not cheap and tend to be used only when there is no effective solid-state substitute, such as indicator lights. A low-cost moving-coil meter might have an accuracy of ±20 per cent; the best and most expensive models will have an accuracy of better than ±1 per cent.

Moving-iron meters

An alternative to the moving-coil meter movement, cheaper to make but less accurate, is the moving-iron meter movement. The basic construction of a moving-iron meter is shown in Figure 6.2. This shows the more common **repulsion moving-iron meter**, based on the fact that two pieces of soft iron are repelled and move away from each other if they are magnetised so that they have the same polarity (either north or south). There is another design, called an attraction moving-iron meter, but most manufacturers use repulsion movements.

When a current flows through the coil, a mag-

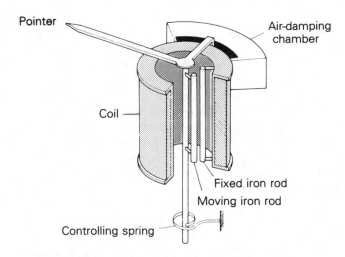

Figure 6.2 Repulsion-type moving-iron meter.

netic field is created by the solenoid and both soft iron rods become magnetised. Clearly, they will both be magnetised in the same direction, so they will repel each other and move the pointer along the scale, against the controlling hairspring. The more current flows through the coil, the more powerful will be the repulsion and the further along the scale the pointer will move.

All moving-coil meter movements are effectively damped – that is, prevented from swinging too rapidly – by magnetic effects. This damping is not present in moving-iron meters to the same extent, so some kind of extra damping, usually a simple air brake, is needed. The air resistance of a small vane in the air-damping chamber (Figure 6.2) prevents the pointer from moving too rapidly along the scale, but does not introduce any long-term effects.

An important difference between the moving-coil meter and the moving-iron meter is that the moving-coil meter involves a magnetic field produced by a permanent magnet and the moving iron meter does not. This means that moving-iron meters can be designed for use with a.c. or d.c., whereas moving-coil meters will work only with d.c.

Shunts and series resistors

Most meter movements (with a few exceptions) are designed to have a high sensitivity and have coils with windings that contain many turns of very fine wire. The resistance of the coil is in the region of a few hundred ohms to several kilohms. A typical FSD might be 1 mA. A meter like this is useless for general-purpose measurement of current, for two reasons.

First, the maximum of 1 mA is too low for the majority of measurements an engineer might want to make. Second, a typical coil resistance of, say, 2 kΩ would seriously affect the current being measured. For example, an e.m.f. of 1 V could drive a current of 10 mA through a 100 Ω resistor. But if we were foolish enough to connect our meter in the circuit in series with the resistor to measure the current flowing, then the current could never be more than 500 μA because of the resistance of the meter itself!

For measuring **current**, in which the meter must be placed in series with the current being measured, the meter should ideally have zero resistance. This is never possible, so a practical current meter should have as low a resistance as is practical. The

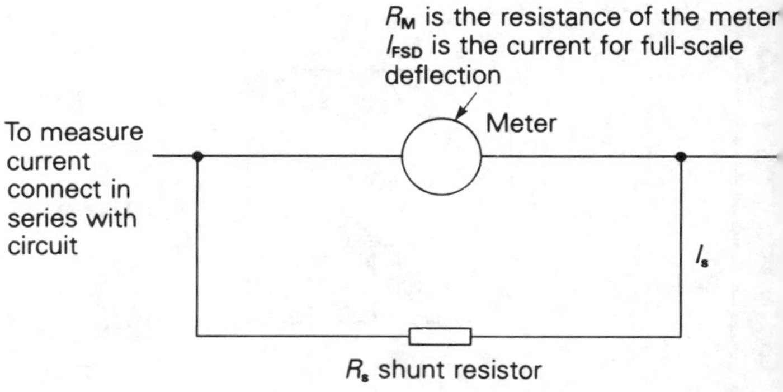

R_M is the resistance of the meter
I_{FSD} is the current for full-scale deflection

Figure 6.3 A shunt resistor, for reading current.

way in which the meter can be 'tailored' to read larger currents, and its resistance reduced, is by the use of a **shunt resistor**, as shown in Figure 6.3. A shunt resistor is simply a resistor connected in parallel with the meter movement. The formula for calculating the value of resistance required for use as a shunt resistor is

$$R_s = \frac{R_M I_{FSD}}{I_S}$$

where R_S is the value of the resistor in ohms, R_M is the resistance of the meter movement in ohms, I_{FSD} is the current required for full-scale deflection of the meter movement (without a shunt) in amps, and I_S is the required current flowing through the shunt, according to the application, in amps.

So our 1 mA FSD meter with a 2 kΩ resistance would need a shunt resistor of 2 Ω if we wanted it to read an FSD current of 1 A. Look at the example above, and you will see that although the meter still affects the measurement it makes much less difference.

For applications where you want to measure **potential difference**, the meter needs to have as high a resistance as possible so that it will put the least possible load across the p.d. When measuring a p.d., the meter is always connected in parallel with the p.d. If a meter is used to read high voltages, a series resistor is fitted (Figure 6.4). The value of the series resistor required is given by

$$R_S = \frac{V_T}{I_{FSD} - R_M}$$

where R_S is the series resistance in ohms, and V_T is the required FSD voltage across the meter plus resistor, in volts.

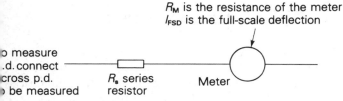

R_M is the resistance of the meter
I_{FSD} is the full-scale deflection

Figure 6.4 A series resistor, for reading p.d.

If we wanted the meter we used as an example, with its 2 kΩ resistance and 1 mA FSD, to measure a p.d. of 100 V, it would need a series resistance of 98 kΩ.

Multimeters

The **multimeter** is one of the most useful items of test equipment that an electrical or electronics engineer can have. Basically, it is a sensitive meter movement, fitted with a whole array of shunts and series resistors that can be selected by means of

Table 6.1 Functions and ranges of a typical multimeter.

Direct voltage	Alternating voltage
0–0.1 V	
0–1 V	
0–10 V	
0–100 V	Ranges as for direct voltage
0–1000 V	
0–5000 V	

Direct current	Alternating current
0–100 μA	
0–1 mA	
0–10 mA	
0–100 mA	Ranges as for direct current
0–500 mA	
0–5 A	

Resistance	
0–100 Ω	
0–1 kΩ	
0–100 kΩ	
0–10 MΩ	

one or more multiposition switches. The range and function controls are made as convenient as possible for ease of use. A good multimeter might have the functions and ranges listed in Table 6.1. Multimeters from different manufacturers will have rather different ranges, but Table 6.1 gives a good idea of what is average in a good-quality instrument.

The input resistance is a factor that is independent of the ranges available; it is the measure of the sensitivity of the meter. A sensitivity of 20 kΩ per volt is typical for the best analogue instruments. Clearly, all else being equal, a meter that loads any p.d. under measurement the least (by putting across it the highest possible resistance) is the best. A meter of high sensitivity will also present the lowest resistance when measuring current.

Table 6.1 includes a range for the measurement of resistance. This is a simple function of the multimeter, using Ohm's law. The multimeter is fitted with a low-voltage source of e.m.f., usually a small battery. To measure resistance, the source of e.m.f. is connected across the resistance to be measured, and the resulting current can be read on the meter scale directly in ohms. The better meters are equipped with a means of stabilising the battery voltage, for accurate resistance readings.

When buying a multimeter for yourself, choose the best you can afford, but remember that you have the option of buying a digital meter (see below).

Measurement of resistance

Multimeters can measure **resistance** directly, and they do this by passing a small direct current through the component being measured. Multimeters are fitted with a battery for this purpose (typically 1.5–9 V). When you select a resistance range, the battery is connected so that it is in series with the meter (which reads current). The amount of resistance between the leads of the meter determines the current that flows, and the meter scale is calibrated according to Ohm's law to show resistance.

The resistance of passive components (like resistors and inductors) can be measured, but **active** components (described in Chapter 7) cannot be reliably measured as they may not follow Ohm's law.

It is very important to make sure that the component being measured is not connected to any power source, as this would not only make the measurement wrong, it could also damage the meter. Measuring a resistor that is in position in a circuit is unlikely to be reliable anyway, as other components in the circuit that are in parallel with the resistor will often make the reading too low.

Digital meters

Modern electronics has come up with an alternative to the analogue meter, in the form of the digital meter. A digital meter is a complex piece of electronics, and it displays a voltage in the form of numbers instead of having a pointer moving along a calibrated scale. Digital meters have many advantages compared with analogue meters, and a few disadvantages.

Advantages

1. Easy to read: the output is in the form of a number, and it is not necessary to look carefully at a scale to determine an exact value.
2. More accurate: price for price, digital meters can be made more accurate than their analogue counterparts.
3. Stronger: the digital meter has no meter movement that can be damaged by hard knocks.
4. Smaller: the most accurate digital meter still only needs a few numbers for its display, whereas an accurate analogue meter needs a large scale so that you can read fine divisions.

Disadvantages

1. Suitable only for constant values. When a current or voltage is steadily changing, an analogue meter will track the changes. A digital display will just be an unreadable blur of numbers.
2. Require a power source: the electronics in a digital meter need a power source, usually a battery, to function.
3. Sometimes misleading: the display may show a value to three decimal places of a volt, but the meter may be far less accurate than it seems!

Both types of instrument find a place in today's workshops.

Insulation testers

It is sometimes necessary for the installations engineer to check that the insulation of a circuit is in order. Since what is under test is the ability of insulation to withstand mains voltages, the testing instrument must check the resistance of the insulation at high voltage. The testing instrument is called an **insulation tester**, although some people will still refer to it as a 'Megger', which is actually a trade name.

Insulation testers apply a high voltage between the points under test (typically between the cores of a mains wiring system) and measure the resistance. The insulation is usually tested at over twice its normal voltage (500 V is applied for 220–240 V wiring).

The first and most popular type of insulation tester – still widely used – has a hand-turned crank that spins a generator to produce the high voltage needed to test insulation. More modern instruments use electronic systems to step up the voltage from a small battery and are used in much the same way as an ordinary multimeter.

Capacitance and inductance meters

Instruments are available that can measure capacitance and inductance directly. It is not very often that the electrical or electronics engineer needs to measure either of these quantities, and many well-equipped workshops have neither of these instruments. Both work by comparing the unknown capacitance or inductance with a reference capacitor or inductor in the instrument or by using the effects of an external capacitance or inductance on a timing circuit.

Capacitance and inductance meters are expensive: perhaps twice as costly as a good multimeter.

Transistor testers

Although transistors (described in Chapter 10) and integrated circuits are at the heart of modern electronic circuits, it is in fact seldom necessary to test a transistor. Transistors are difficult to test thoroughly, and are relatively cheap, so it is normal to replace any device that is suspect. It is, however, possible to buy instruments that will measure all a transistor's parameters accurately.

Somewhat more useful for most engineers are in-circuit transistor testers that can perform a 'go/no go' test on a transistor without removing it from the circuit in which it is fitted. Such devices are useful servicing aids.

The oscilloscope

The **cathode-ray oscilloscope** (CRO for short) is probably one of the most useful items of test equipment to be found in the workshop, after the multimeter. The CRO uses a **cathode-ray tube** to display the waveform of an electrical voltage directly on a screen.

The cathode-ray tube

Most of us are familiar with the appearance of the cathode-ray tube (CRT) from looking at the screen of a television, which is also a CRT. Figure 6.5 shows the basic design of the CRT.

The **cathode** is just like a lamp filament, though it is treated with thorium to increase the number of electrons that can be driven off when it is heated.

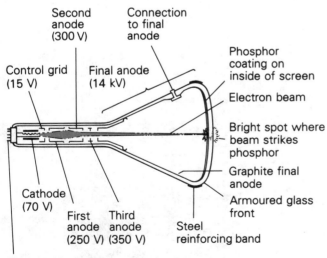

Figure 6.5 A cathode-ray tube from an oscilloscope.

Labels in figure:
Second anode (300 V)
Connection to final anode
Control grid (15 V)
Final anode (14 kV)
Phosphor coating on inside of screen
Electron beam
Bright spot where beam strikes phosphor
Graphite final anode
Armoured glass front
Cathode (70 V)
First anode (250 V)
Third anode (350 V)
Steel reinforcing band
Tube base: pins connect to heater, cathode and all anodes except the final anode

Facing the cathode is the **final anode**, a metal plate connected to an outside terminal. The two parts are sealed inside a glass envelope in which there is a vacuum.

In front of the cathode is a cylindrical **control grid**, and in front of the control grid are two cylindrical **anodes**, carefully designed to act as a 'lens' when the proper voltages are applied. The object of these anodes (termed **first** and **second anodes**) is to concentrate the electrons into a beam, aimed at the front of the tube. Figure 6.6 shows the way that the electric field produced by the two cylindrical anodes acts as an electron lens to concentrate the beam of the electrons, just like a glass lens can be used to concentrate a beam of light.

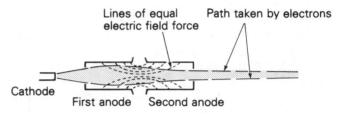

Labels in figure:
Lines of equal electric field force
Path taken by electrons
Cathode
First anode
Second anode

Figure 6.6 The electric field between the first and second anodes concentrates the beam of electrons, in much the same way that a lens concentrates light.

Next there is a third anode (usually called the **focus anode**); voltage applied to this can be finely adjusted to bring the electron beam to a sharp focus on the screen. Lastly, there is a final anode, which is connected to a graphite coating applied to the inside of the flared part of the tube. All the flared part of the tube is thus at the same potential as that of the final anode; in fact the whole of the front part of the tube could be said to be the final anode, as it is all connected together.

The front part of the tube – the screen – is coated on the inside with a material called a **phosphor**. This has the property of glowing brightly when struck by electrons (phosphors are also used on the inside of fluorescent lamps to make them glow). Phosphors can be made almost any colour; a monochrome television tube would use one that glows white, while oscilloscopes and computer monitors often have a green screen.

Operation of the tube is straightforward in principle. Electrons are emitted by the heated cathode, formed into a beam by the first two anodes, and focused by the third anode. The beam strikes the phosphor on the screen to produce a bright dot.

The beam current, and thus the brightness of the dot, can be controlled by the voltage applied to the control grid.

The tube is quite long, so the electrons need a lot of energy to persuade them to travel down it. Moreover, they must have enough energy left to make the phosphor glow brightly when they reach the screen. For these reasons, the voltages associated with CRTs are high: so high, in fact, that the final anode connections cannot be made to the pin connectors at the back of the tube; this would pose insurmountable insulation problems. Instead, the high-voltage connection is made to a special plug on the side of the flared part of the tube.

Beam deflection

So far we have described a CRT that produces a single dot in the centre of the screen. The brightness can be controlled by the grid, but we have to use some extra parts to move the dot around the screen.

The oscilloscope is a measuring instrument that displays electrical waveforms on a CRT, the screen being marked with a grid representing time and voltage. An accurate and linear method of beam deflection is needed, and **electrostatic deflection** is the answer. The CRT is fitted internally with parallel horizontal and vertical plates, mounted in front of the focus anode. A high voltage applied to opposite plates will deflect the electron beam, and the degree of deflection on the screen is proportional to the applied voltage.

Oscilloscope controls

The oscilloscope screen is ruled with a grid of squares, called a **graticule**, and is arranged like a graph, with a vertical y axis. A typical CRO screen layout is shown in Figure 6.7.

The grid is conveniently ten squares in both directions. The (horizontal) x axis is calibrated in time, and the (vertical) y axis is calibrated in voltage. Both the time scale and the voltage scale are adjustable by means of range-setting controls. A small general-purpose oscilloscope is shown in Figure 6.8. The x axis is calibrated in 'time per division', making the spot scan the screen at a rate controllable from a maximum speed of 1 μs per division to a minimum of 0.1 s per division (in the model shown). The range-setting control is at the top right of the front panel. In the centre of the

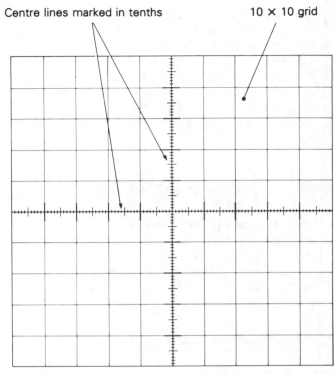

Centre lines marked in tenths   10 × 10 grid

Figure 6.7 A typical oscilloscope screen graticule.

lower part of the panel is the control for setting the range of the y axis. This is variable, in this particular model, in 12 steps from a maximum sensitivity of 5 mV per division to a minimum sensitivity of 20 V per division. The control is in the middle of the lower part of the photograph.

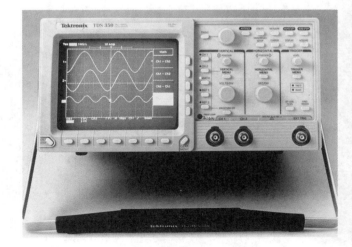

Figure 6.8 A typical small oscilloscope.

Oscilloscopes characteristically have a very high input resistance, corresponding to a sensitivity that is better than most multimeters. A graph of almost any electrical waveform can be displayed on the screen, providing an invaluable aid to understanding circuit function or to servicing most electronic equipment.

A most important feature to be found on the oscilloscope is the trigger or synchronisation system. Circuits in the instrument detect the level of the signal being measured and trigger the sweep of the x axis (at the set rate) at a predetermined point on the waveform. Because the sweep is always triggered at the same point on the waveform, the picture on the screen is automatically 'locked' in place on the screen and does not drift from side to side with small changes in frequency. Provision may be made to sweep the x axis from an external source, perhaps derived from the circuit under measurement. Additionally, there may be special-purpose filter circuits built into the CRO to select (for example) television synchronisation signals, and lock the picture onto them. A range of such facilities can be seen on the instrument in Figure 6.8.

The CRO can be used as a measuring instrument, since the 'graph' on the screen is accurately calibrated. Voltage can be measured with reference to the y axis, and frequency or pulse widths by reference to the x axis. The more expensive models have a double-beam tube, which displays two waveforms simultaneously. The x axes are synchronised with each other but the y axes are controlled by independent inputs. This facility makes it possible to compare two different waveforms.

Signal generators

The signal generator is an instrument for producing waveforms of known shape at a known frequency. Most signal generators can produce a sine-wave or square-wave output, with a peak-to-peak voltage ranging from a few millivolts to a few volts. Frequencies from a few Hz to a few MHz are standard in an inexpensive instrument.

Frequency meters

Meters are available that will measure frequency very accurately and show the result on a digital display. They are rather expensive and are not widely used because the oscilloscope can be used to measure frequency to a moderate degree of accuracy.

■ CHECK YOUR UNDERSTANDING

● **Safety** is important in making measurements: your own, that of the measuring instrument, and that of the circuit or system being measured.

● Knowing **how** accurate a measurement needs to be is as important as the **accuracy** of a measurement.

● **Moving-coil** meter movements are more often used than moving-iron movements. **Shunt** (parallel) and **series** resistors are used to make the basic meter read a range of currents and voltages.

● **Digital** meters indicate the reading on a digital display. They are accurate and generally more robust than **analogue** meters.

● Digital meters can be used only when voltages or currents are unchanging, or are changing very slowly.

● An **insulation tester** (Megger) should be used to check wiring that is to carry mains voltage.

● An **oscilloscope** is an instrument – rather like a small television – that displays a graph of an alternating signal, plotting voltage (vertical axis) against time (horizontal axis) on a screen marked with a ruled graticule.

REVISION EXERCISES AND QUESTIONS

1 Typically, what is the accuracy of a moving coil multimeter?
2 Indicate the principal advantages and disadvantages of a digital multimeter, compared with a moving-coil analogue meter.
3 When you use an oscilloscope to examine an audio signal, what does the instrument tell you about the signal (at least three things)?
4 What is a frequency generator for?
5 State *two* measures of the sensitivity of a moving-coil meter.

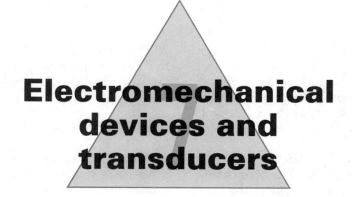

Electromechanical devices and transducers

Introduction

This short chapter deals with a range of passive and 'non-electronic' devices used in electronics and power engineering, describing them briefly and outlining their functions.

Electromagnetic devices

Relays

Relays were among the first electrical components used as amplifiers, and were widely applied to early telegraph networks. A relay consists of an electromagnet that operates some sort of mechanical switch. A low voltage and current can operate the relay and cause the switch to close, the switch controlling a larger current or voltage. Relays can therefore be used as amplifiers; this was one of

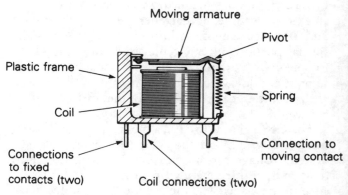

Figure 7.1 A small changeover relay.

their first uses, routeing and amplifying telegraph signals. A simple relay is illustrated in Figure 7.1.

In a telegraph system, in which pulses of current are sent at a low rate of repetition, a relay can be used to amplify signals. The principle is shown in Figure 7.2.

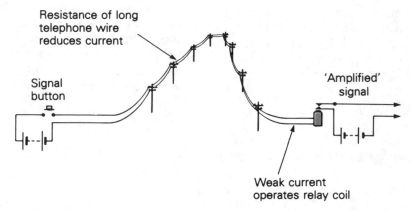

Figure 7.2 A relay used as an amplifier in a telegraph system.

There is no reason why the relay should not work more than one contact, nor is there any reason why contacts should not be normally closed (and open when the electromagnet is energised) rather than the more usual 'normally open' form. Various circuit symbols are used for relays; Figure 7.3 shows two of them. Both relays are shown with two pairs of contacts, one normally open and one normally closed. It is conventional to draw the contacts in the positions they adopt with the coil energised.

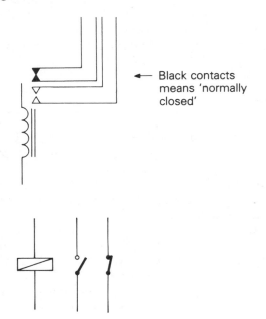

← Black contacts means 'normally closed'

Figure 7.3 Circuit diagrams used for delays.

It is not always convenient to represent the relay contacts as neatly as in Figure 7.3, as a relay may have several sets of contacts controlling widely separated circuits; in this case the contacts are drawn in a convenient position, and are clearly labelled as belonging to a particular relay coil.

Relays come in an amazing range of sizes. At one extreme, there are huge machines in power stations and distribution systems, switching thousands of amps at thousands of volts. At the other end of the scale there are tiny low-current relays encapsulated in tiny metal cans no bigger than your little fingernail.

Reed switches

The magnetic reed relay was developed for telecommunications before solid-state switching became possible. It is a cheap and extremely reliable method of switching a low-power signal.

The **reed switch** consists of two springy blades of ferrous metal, sealed in a glass envelope full of inert gas. The blades are mounted so that they are not quite touching, about 0.5 mm apart. A reed switch is shown in Figure 7.4. If a magnetic field is brought close to the reed switch, the reeds become temporarily magnetised and stick together, completing the circuit. When the magnetic field is removed, they spring apart again and break the connection.

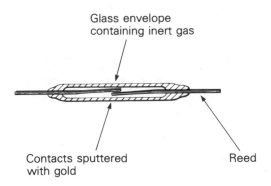

Glass envelope containing inert gas

Contacts sputtered with gold

Reed

Figure 7.4 A 'single make' reed switch.

Reed switches are often used in security alarms to protect doors (and windows). The reed switch – wired into a suitable alarm system – is fixed to the frame, and a permanent magnet is fixed to the door so that it is next to the reed switch when the door is shut. Opening the door moves the magnet away from the reed switch, which opens and sounds the alarm.

Reed relays

A reed relay is made from a reed switch simply by winding a coil round the outside of the reed switch. When a current flows through the coil, the induced magnetism makes the reed switch close.

The ends of the reed are coated with gold, and this, combined with the inert gas, prevents the contacts from corroding. The reed relay is very reliable, and is fast in operation, closing in about 1–2 ms, and opening in less than 0.5 ms. Reed relays should not in general be used with inductive circuits, or arcing will damage the thin gold layer and may even weld the contacts shut.

Changeover reed switches are also common, and are constructed as shown in Figure 7.5.

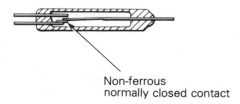

Non-ferrous
normally closed contact

Figure 7.5 A changeover reed switch.

Reed relays are commonly encapsulated in DIL packs. DIL stands for 'dual in line' and refers to the arrangement of the connecting pins along the sides of the plastic pack. The pins are exactly one tenth of an inch apart. DIL packs are commonly used to encapsulated **integrated circuits** (Chapter 12), which makes them ideal for use on **printed circuits** (Chapter 17). There are still occasions when a relay is the best answer to a design problem, providing as it does complete isolation between circuits, and a capability of switching alternating or direct currents. Two DIL reed-relay pin diagrams are given in Figure 7.6.

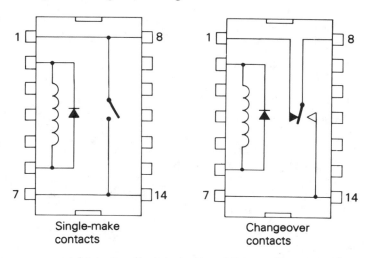

Single-make
contacts

Changeover
contacts

Figure 7.6 Pin-connection diagrams for two types of DIL reed relay.

Solenoids

Where it is required to change electrical power into linear mechanical motion, a **solenoid** may be used. It is simply a coil wound round a suitable former, with a soft iron core free to move inside. When the coil is energised, the core is pulled into the coil by the magnetic field produced by the current flow. Solenoids are used to operate the mechanical parts of some cassette tape transport mechanisms, for instance. Solenoids are not particularly efficient in

their use of power, and a large current is required if useful mechanical work is to be done. Small permanent-magnet motors are more efficient.

Transducers

The division between electromechanical devices and **transducers** is a blurred one for, in a sense, any machine that turns an electrical signal into motion, sound, etc. (or the other way round) is a transducer. Relays and solenoids are not generally classed as transducers, however. The devices in this section are.

Speakers

A speaker (or loudspeaker) is, in essence, a special kind of solenoid. A typical speaker is illustrated in Figure 7.7. A powerful **magnet** surrounds the **speech coil**, which is connected to the amplifier, etc., driving the speaker. A current flowing through the coil in one direction pulls the coil back into the magnet, whereas a current flowing through the coil the other way will push the coil out of the magnet. When the speaker is connected to the output of an audio amplifier, the movements of the coil will duplicate the waveform of the amplifier's output.

In order to produce a loud sound output, the coil has to move as much air as possible, so

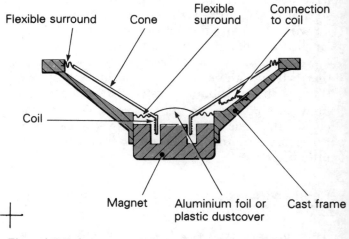

Flexible surround   Cone   Flexible surround   Connection to coil

Coil

Magnet   Aluminium foil or plastic dustcover   Cast frame

Figure 7.7 A cross-section through a speaker.

the **speaker cone** is made as large as is practical. The movement of the cone – which has to be rigidly fixed to the coil – causes vibrations in the air (alternate compressions and rarefactions), which reproduce the sounds fed into the audio system.

The design of speakers has been given a great amount of attention in the search for a speaker that will give the highest possible fidelity. Although a small speaker can reproduce the audio spectrum (the range of sounds that are audible to humans) reasonably well, it cannot provide the best possible fidelity across the whole range. Hi-fi systems therefore use two or three speakers, along with a simple electrical filter system to split the signal between the two or three speakers, according to frequency.

A bass speaker, for example (known as a 'woofer' in the hi-fi world), must move a lot of air to reproduce low bass sounds with sufficient volume. A large cone is necessary, or if the size requirement of the speaker does not allow this, then a small cone that has a large excursion (i.e. it moves in and out a long way) will suffice.

A treble speaker (known as a 'tweeter') does not have to move as much air, but in order to reproduce high-frequency notes the cone must move very quickly. The large – and consequently heavier – cone of the bass speaker has too much inertia to move rapidly backwards and forwards, and so could not be used as a tweeter. Tweeters have small, very lightweight cones, with relatively little travel.

Since the output power of an audio amplifier has to be dissipated as heat and sound by the speech coil, there are limits on how much power a given speaker can handle.

Piezoelectric sounders

Where a quiet 'bleep' is required, rather than an undistorted speech or music output, piezoelectric sounders may be used instead of small speakers. Applications are computer sound output systems, keyboard bleepers, small alarm bleepers and electronic alarm watches. The piezoelectric sounder is no more than a small piece of piezoelectric crystal, sometimes with and sometimes without an extra 'cone' (which, being flat, is referred to as a diaphragm). Low-voltage electrical signals applied to the crystal will make it flex, so a suitable audio frequency waveform will cause an audible output. Piezoelectric sounders are very small, cheap, and not very loud. There is a diagram of the way a piezoelectric sounder works in Figure 2.7 of Chapter 2.

Microphones

Crystal microphones

There are three main types of microphone in common use: crystal, moving coil and capacitor. The cheapest is the crystal microphone, which is very similar in design to the piezoelectric sounder. A simplified drawing of a crystal microphone is given in Figure 7.8. Sound waves hitting the microphone will move the diaphragm in sympathy with them, flexing the diaphragm. The diaphragm moves the piezoelectric crystal, which generates an electrical output that is an analogue of the sound waves.

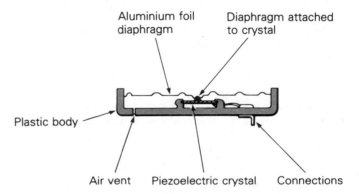

Figure 7.8 A crystal microphone.

Crystal microphones characteristically have high voltage outputs: relatively speaking, for the crystal microphone may produce a few millivolts. They also have a very high internal impedance – several megohms – and so are not very suitable for use with bipolar transistor amplifiers.

Another disadvantage is that the output of the crystal microphone is not particularly linear (some sound frequencies produce more electrical output than others), giving poor sound quality. The crystal microphone has therefore been replaced in most common uses by moving-coil or electret microphones.

Moving-coil microphones

Moving-coil microphones are made in a wide range of sizes and prices, from cheap models supplied with inexpensive cassette recorders to more expensive 'studio' microphones that have a very high audio quality. Figure 7.9 shows a simplified drawing of a moving-coil microphone. The construction is not unlike that of a small speaker. Sound waves move the **diaphragm**, to which is fixed a lightweight **coil**. The coil is surrounded by a magnetic field, so as it moves a current is induced

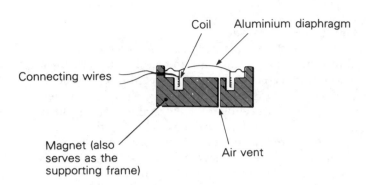

Figure 7.9 A moving-coil microphone.

in the coil. The size of the current accurately mirrors the frequency and amplitude of the sound waves.

The coil impedance decides the impedance of the microphone. There are physical limits on how many turns of wire can be used, and for this reason most moving-coil microphones have a standard impedance of 200 Ω. This is suitable for matching both bipolar and FET amplifiers. The current output of the moving-coil microphone is small: a millivolt at most. Moving-coil microphones are cheap and efficient, and can give good sound.

Electret microphones

The most recent addition to the range of commonly used microphones is the **capacitor microphone**, which recent advances have brought out of the broadcast studio and into even the cheapest cassette tape-recorders.

The capacitor microphone utilises an **electret** element, which is a special capacitor with one 'plate' being a flexible diaphragm. If a voltage is applied to the electret element the capacitor becomes charged, and there is a fixed p.d. across the terminals. No current flows, apart from a negligible leakage current.

Sound waves hitting the diaphragm cause it to vibrate, which changes the spacing between the plates of the capacitor in sympathy with the sound. This alters the capacitance (which depends, you may remember, on the spacing between plates). The amount of charge on the capacitor has to remain the same (there is nowhere for it to go), and so as the capacitance varies the p.d. across the capacitor varies to accommodate a constant amount of charge. The change is very small, so an integral IGFET preamplifier is fitted (see Chapter 11), to provide a current output and a typical impedance of 600 Ω.

Electret capacitor microphones give a high output compared with other types, and the sound quality is excellent. They are also cheap and easy to make. It is necessary to provide the electret element with a polarising voltage, to charge the capacitor; this is conveniently done with a 1.5 V battery, an MN1500 cell lasting for several years. The smallest electret microphones are very tiny, powered with button-sized mercury or silver-zinc cells. A simplified diagram of a capacitor microphone is given in Figure 7.10.

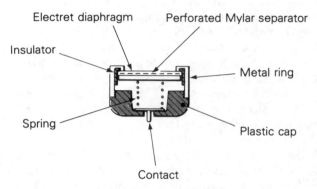

Figure 7.10 An electret capacitor microphone.

Light-dependent resistors

All semiconductors are sensitive to light, to a greater or lesser extent. It is for this reason that transistors and diodes are packaged in light-proof encapsulations, either metal or plastic.

Certain components make use of light sensitivity, and are designed to be used as light-detecting devices. There are several different basic devices, but the simplest and one of the most commonly used is the **light-dependent resistor** (LDR), also known as the **photoconductive cell**, or **photoresistor**. LDRs make use of the fact that cadmium sulphide (chemical symbol CdS) is particularly sensitive to light in the visible part of the spectrum.

In principle the CdS photoresistor is no more than a thin layer of cadmium sulphide on a ceramic base, with metal (usually aluminium) connections printed on top. The electrodes are made with a characteristic interlocking comb layout, to maximise the length of the 'junction' in relation to its width. Figure 7.11 illustrates the design.

In the dark, the CdS photoresistor has a high resistance. For instance the ORP12, the standard 'large' photoresistor, has a **dark resistance** of 10 MΩ. In bright sunshine the resistance can drop

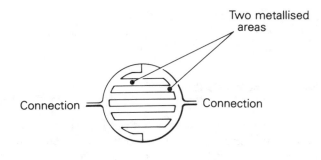

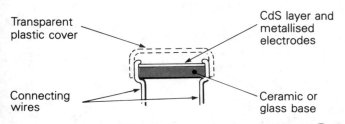

Figure 7.11 A typical photoresistor, or LDR (CdS̄ cell).

accounts for the rather small number of different designs produced by the different manufacturers. The ORP12 and the smaller ORP60 take care of most requirements.

CdS photoresistors can dissipate moderate amounts of power: for example the ORP12 can dissipate up to 200 mW. They can also be used with relatively high voltages; the ORP12 can survive 110 volts maximum. Because photoresistors are symmetrical, they are unaffected by polarity of applied voltage and can be used with a.c. or d.c.

Figure 7.12 illustrates the CdS photoresistor in a typical application, turning a small lamp on when it gets dark and off again at dawn. The circuit symbol is shown with three arrows denoting the 'light'.

The circuit of Figure 7.12 is completely practical and can be made as a project, but not until you have studied some of the other components used in this circuit; it involves **diodes** (Chapter 9), **transistors** (Chapter 10), and a knowledge of **construction methods** (Chapter 17).

Although photoresistors have many uses, they respond to changes in light rather slowly, and so are unsuitable for remote control applications, or other applications where a rapid response to changes in light level is needed.

to as low as 150 Ω or even less. Compared with the small changes we are used to dealing with in electronics this is a massive range, which probably

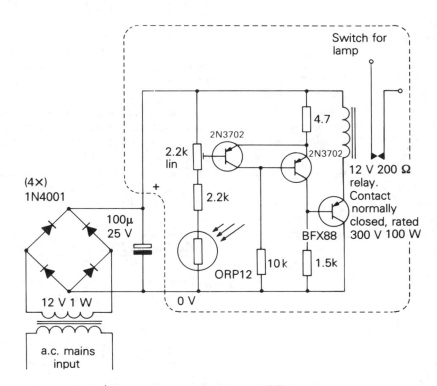

Figure 7.12 *A lamp controlled by a LDR to come on when it gets dark.

Thermistors

Used in temperature-sensing applications, **thermistors** are semiconducting resistors which change resistance with temperature. There are two types: **negative-resistance temperature coefficient (NTC)**, in which the resistance decreases as temperature increases; and **positive-resistance temperature coefficient (PTC)**, in which the resistance increases with increasing temperature. Circuit symbols are given in Figure 7.13.

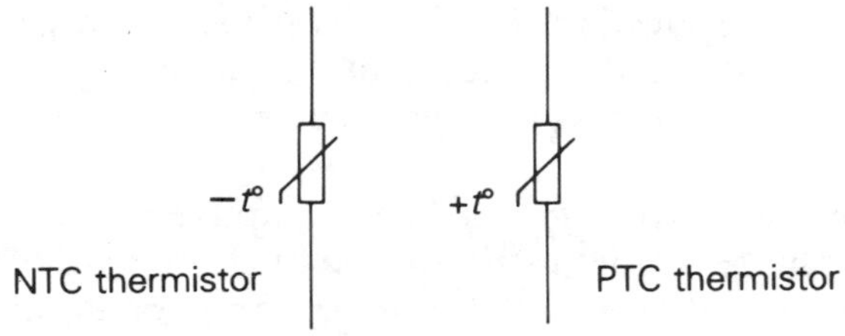

NTC thermistor   PTC thermistor

Figure 7.13 Thermistors.

Like 'ordinary' resistors, thermistors come in a range of physical size, resistance, and power dissipation. Their characteristics (rate of change of resistance with temperature) also vary from one to another, and manufacturers usually supply a graph of resistance against temperature.

Voltage-dependent resistors

Voltage-dependent resistors (VDRs) are used in only a few applications. A VDR is simply a resistor whose resistance **decreases** as the voltage across the resistance increases. VDRs can be made with a high resistance and high operating voltage, so they are useful in high-voltage systems where the more usual semiconductor devices cannot easily be used. The circuit symbol for a VDR is given in Figure 7.14. A VDR connected across a power supply will provide a measure of protection from transient high-voltage peaks. A VDR can be chosen to have a very high resistance at the supply voltage, but a

Figure 7.14 Circuit symbol for a voltage-dependent resistor (VDR).

low resistance at the high voltage levels associated with transients that might cause damage or cause interference to the circuits the VDR is protecting.

Hall-effect devices

Named after the discoverer of the fact that magnetism could affect charge carriers (holes and electrons; see Chapter 8) in a solid, Hall-effect devices are semiconductors that react to an external magnetic field. Hall-effect devices are sensitive not only to the existence of a magnetic field but also to its polarity.

Hall-effect devices are used for the measurement of magnetic fields, and also as switches, where they make a useful alternative to mechanical and optical switching systems.

▮ CHECK YOUR UNDERSTANDING

● A **relay** is a switch – usually mechanical – that is operated by an electromagnet. A relay can use a small current to turn a large current on and off, or a low voltage to control a high voltage.
● A **transducer** is a device that turns an electrical signal into sound, movement, etc., or vice versa.
● **Speakers** (sometimes called 'loudspeakers') change an electrical signal into sound at audio frequencies.
● **Microphones** change sound into an electrical signal. Only two types are widely used today: the **moving-coil** microphone and the **electret** microphone.
● **Light-dependent resistors** change resistance according to the amount of light falling on them.
● **Voltage-dependent resistors** change resistance according to the voltage across them.
● **Hall-effect devices** change conductivity according to the strength and direction of the surrounding magnetic field.

REVISION EXERCISES AND QUESTIONS

1 Draw a sketch of a simple changeover relay.
2 Describe how a speaker converts an electrical signal into sound.
3 Draw the circuit diagram for a light-dependent resistor, and describe what it does.
4 Telephones almost always use moving-coil microphones. Why do you think this type is used in preference to crystal or electret microphones in this application?

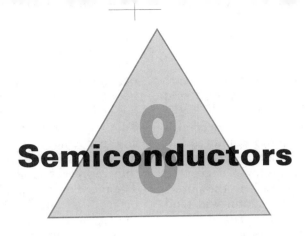

Semiconductors

Introduction

We are now going to have a look at atomic physics. If you are to understand how modern semiconductor devices – from transistors to microprocessors – work, you need to know a little about the physics of their operation. The problem with atomic physics is that most of it goes on at a level that is far too small to see, even with the most powerful microscopes. Not only this, but the 'common sense' rules that apply to the world around us don't necessarily apply to the world of atomic physics. We therefore have to rely on what are known as **models** of reality, and doing this (as you will discover later on) sometimes makes things appear rather puzzling.

Modelling the atom

In the section of Chapter 2 dealing with electricity, I used a model of the atom originally designed by Niels Bohr. It is quite easy to picture the Bohr atom, with its hard, bullet-like electrons hurtling round the massive nucleus just like planets orbiting a sun in a tiny solar system. But remember – and forgive me if I keep repeating this – we are considering **models** of atoms, and not the real thing.

The work done by Werner Heisenberg in the late 1920s showed that Bohr's model is unfortunately further from reality than we might hope; an atom is actually a rather fuzzy and uncertain thing, not at all like Bohr's microminiature solar system. This is not a book about atomic physics, so it is unnecessary to look too closely at the construction of atoms, but it is important to realise that there are 'rules' that appear to govern the behaviour of atoms and their component parts. Many of these rules seem contrary to what we would expect, but

our feelings are based on what we know about the behaviour of objects that are much larger than atoms and electrons.

Let us begin by considering a single atom of an element: silicon is a useful example. The atom consists of a central nucleus surrounded by a cloud of electrons, which can be represented diagrammatically as in Figure 8.1. The electrons arrange themselves into three orbits, or 'shells'. Although the diagram shows the electrons in a flat plane, the orbits actually occupy a spherical **shell**. The shells are given letters, starting with K for the innermost shell, then L and M, and if there are more than three shells N, and so on. Each shell can hold a specific number of electrons; two in the K shell, eight in the L shell, and 18 in the M shell. As we build up models of different atoms, the shells are filled from the orbit nearest the nucleus, so silicon (which has 14 electrons) has full K and L shells (two and eight) and the remaining four electrons in the M shell.

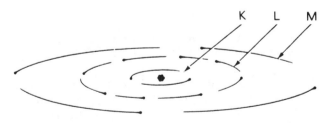

Figure 8.1 A simple model of a silicon atom.

This seems straightforward enough, but if we look closer at the atom we find a little of Heisenberg's 'fuzziness' beginning to creep in. One of the rules governing the behaviour of electrons in a system (a 'system' means an atom or groups of atoms) states that no two electrons can be at precisely the same energy level.

What does 'energy level' mean in this context? Think back to the solar system analogy and imagine a spacecraft orbiting in the K shell. If the pilot runs the engines to increase the spacecraft's speed, it will move out to a more distant orbit, perhaps even as far as the L shell. Subtract energy from the spacecraft by allowing some energy to dissipate as heat (if it's in a low orbit in the outermost fringes of the atmosphere, friction will tend to slow it) and it will drop into a lower orbit. Thus it is clear that the greater the energy possessed by the spacecraft, the higher – further from the nucleus – will be its orbit.

If no two electrons are allowed to have the same energy, it follows that no two electrons can orbit at exactly the same distance from the nucleus. It also follows that all the shells consist of more than one possible orbit. The L shell, with its eight electrons, must consist of at least eight different orbits, all close to one another but not the same. Can we say how these orbits are arranged, and which of the possible orbits are in fact occupied by an electron? Unfortunately we can't. Another of the rules governing the behaviour of electrons says that we cannot know the speed, position and direction of an electron all at once. We can never say for certain the whereabouts of the eight electrons that form the L shell at any particular instant; all we can do is say where there is the greatest probability of their being located. Compare part of the 'Bohr' orbit in Figure 8.2(a) with the 'Heisenberg' version in Figure 8.2(b).

(a)   (b)

Figure 8.2 Two models showing sections of an electron's orbit: (a) Bohr; (b) Heisenberg.

The darkness of the shading in Figure 6.2(b) represents the **probability** of an electron being in that particular orbit: the darker the shading, the more likely an electron is to be found there. 'Probably' is all we can say about the electrons, not because of any limitations in our measuring equipment, but because of the very nature of electrons. This rather puzzling fact about electrons is one of the more important discoveries to come out of modern quantum physics.

Bohr's model is reliable in saying that there are specific regions (or shells) that the electron can

occupy. It is not possible for electrons to orbit between the shells. The areas they cannot occupy are called **forbidden gaps**. We can redraw Figure 8.1 to show the probabilities of electrons being in any particular orbits (Figure 8.3). It now seems more realistic to call the shells 'energy bands', as they represent a range of possible electron energies. The more energy an electron has, the further it will be from the nucleus. Electrons may not orbit in the forbidden gaps between the shells. The outer band is called the **valence band** and is the only band that may not be completely full.

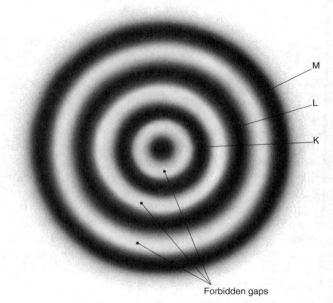

Forbidden gaps

Figure 8.3 An atomic model showing energy bands and forbidden gaps.

If we take a section through this model of the atom, from the nucleus to the outermost band, we can represent this by a diagram showing just a small part of each orbit, like the one in Figure 8.4. We can use this as an **energy-level diagram** for our atom, since the bands represent electrons having increasing energy as we go up the diagram. The degree of 'fuzziness' of the bands depends on the temperature, and it is normal to show energy-level diagrams for a temperature (at or very near) absolute zero, when they are at their least fuzzy.

It is possible to make a leap of the imagination and use one energy diagram to represent the **average** of a very large number of atoms in a system. This is what we do to describe the operation of semiconductor devices. Such energy-level diagrams are, as we shall see, very useful aids to understanding quite subtle atomic interactions. But we can simplify the diagram a little before using it.

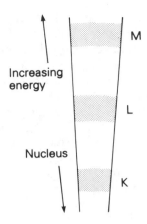

Figure 8.4 A section of the orbits shown in Figure 8.3.

None of the inner shells of atoms are generally involved in any of the interactions that occur in electronics. This means that we can omit the lower bands. Figure 8.5 shows an energy-level diagram for a piece of silicon, just showing the valence band.

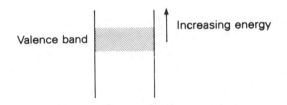

Figure 8.5 Possible orbits in the valence band.

There are a vast number of possible orbits – all different – making up the band. Many or most of these possible orbits will be unoccupied. It is quite possible to imagine a whole band that is completely unoccupied! Such a band is still there (theoretically at least) as it defines the probable positions of any electrons that somehow gain so much energy that they leave the valence band. Because such electrons form an important part of the interactions that take

place in electronics, it is useful to add another band to the energy-level diagram, an 'empty' band beyond the valence band: this band is called the **conduction band**. Both it and the valence band are shown in Figure 8.6.

Conductors and insulators

We can use energy-level diagrams to explain why conductors conduct and insulators don't. Compare the energy-level diagrams for copper and sulphur (Figure 8.7) . In Figure 8.7(a) there is no forbidden gap between the valence and conduction bands. Electrons can move freely from band to band and there is no barrier to electron movement; a small increase in energy can move an electron into the conduction band. Once in the conduction band, the electrons are not bound to the structure of the atoms, and are free to drift through the material as an electric current.

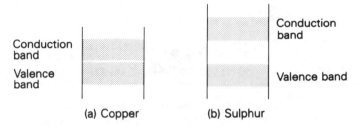

Figure 8.7 Energy-level diagrams of (a) a typical conductor (copper) and (b) a typical insulator (sulphur).

Sulphur is a different kind of material. The sulphur atom has a large forbidden gap between the valence and conduction bands. A moderate increase in energy will not be sufficient to move an electron out into the conduction band, but will only lift the electron as far as the forbidden gap. Since electrons cannot orbit in the forbidden gap, they will not be able to accept such an energy increment, and will stay in the valence band. The conduction band will remain empty, and no current can flow.

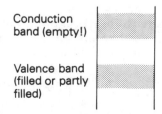

Figure 8.6 Energy-level diagram showing the valence and conduction bands. The conduction band need not have any electrons in it.

Intrinsic semiconductors

Figure 8.8 shows an energy-level diagram for **germanium**. The forbidden gap between the valence and conduction bands is very small and it requires only a little added energy, such as the thermal energy available at normal room temperature, to cause a few electrons to jump into the conduction band. Germanium will therefore conduct electricity, but poorly: less than a thousandth as well as copper. Germanium is called an **intrinsic semiconductor**, since it naturally has properties between those of conductors and those of insulators. The conductivity of germanium is strongly affected by temperature, as you might expect: the higher the temperature, the more the energy bands blur and expand, and the smaller the forbidden gap becomes.

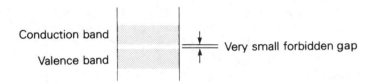

Figure 8.8 Energy-level diagram showing a typical intrinsic semiconductor.

Charge carriers in semiconductors

We have seen how electrons can drift through a material in the conduction band and how this constitutes a flow of electric current. Electrons that are free to move about in this way are called **charge carriers**, because they carry electric charge through the material. But there is another mechanism by which charge can be carried through a substance, and we can see how this operates by looking at an energy-level diagram showing what happens when an electron makes the transition from the valence band to the conduction band (see Figure 8.9).

The movement of an electron from the valence to the conduction band puts a 'free' electron into the conduction band, but it also leaves a **hole** in the valence band: a gap into which an electron might easily move. The hole could be filled by another electron in the valence band, but this would leave a hole somewhere else, which could be filled by another electron, leaving another hole in the valence band, and so on . . .

The creation of the **electron–hole pair** that results from an electron moving from the valence to the conduction band actually allows electrons to move about freely in the valence band as well as in the conduction band, though it is conventional to think of movement of electrons in the conduction band, and of movement of holes in the valence band. Figure 8.10 shows clearly how a movement of holes is really the same as a movement of electrons in the opposite direction; both ways of looking at it are actually valid. Just as the electron carries one unit of negative charge, so the hole can be said to carry one unit of positive charge, which will in the right circumstances exactly cancel out the negative charge on the electron.

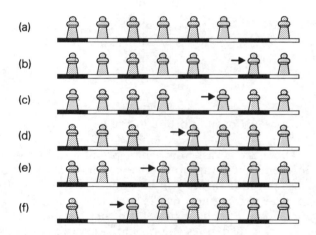

Figure 8.10 Movement of electrons in one direction is the same as movement of holes in the other direction; this can be demonstrated using a draughts (chequers) board with a row of seven pieces. (a) shows the initial arrangement, and one piece moves to the right in each stage of this diagram, (b) to (f). Although no electron has moved more than one place, the hole has drifted all the way across the board.

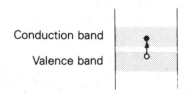

Figure 8.9 When the electron moves from the valence band into the conduction band, it leaves a hole in the valence band.

The idea of a hole moving and carrying a positive charge is a bit odd, and a reasonable person might ask at this stage just why we should bother to describe such a concept when the result is simply an electron drift through the conduction and valence bands. The answer is that in the artificial semiconductors used in electronic devices an excess of either holes or electrons is introduced into a substance, and it is often one or the other mechanism that dominates the conduction inside the material. Artificial semiconductors – materials in which holes or electrons have been deliberately added to the atomic structure – are known as **extrinsic semiconductors**.

Extrinsic semiconductors

Making a semiconductor is easy: in theory, at least. In practice, the technology of making semiconductors has taken two decades to get to its present standard of near-perfection. We start with a cheap, plentiful semiconductor like silicon, purify it until it is absolutely pure, and then add a tiny amount of another substance: one part in 100 000 000 of **phosphorus**, for example. The resulting mixture is then made into a single perfect crystal.

The atomic structure of the phosphorus atom fits nicely into the crystalline matrix of the silicon, but with one electron left over in its outer shell, for it has five and not four electrons in its valence bands. This extra electron is 'spare' to the structure of the crystal, and moves easily into the conduction band. An energy-level diagram for a crystal of silicon 'doped' with phosphorus is shown in Figure 8.11. We show the extra electrons in the conduction band as minus signs. Such a doped semiconductor is called an **n-type semiconductor**, the n representing 'negative' to indicate that the material has extra (negatively charged) electrons.

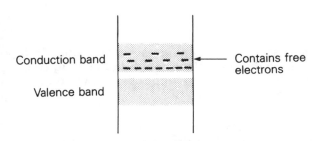

Figure 8.11 An energy-level diagram for an n-type semiconductor.

Alternatively we can add to the silicon crystal matrix a few parts per million of **boron**. The boron atom has only three electrons in its valence band, and it too drops neatly into the silicon crystal structure, but it leaves a hole. This hole is easily

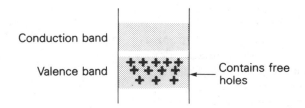

Figure 8.12 An energy-level diagram for a p-type semiconductor.

filled by an electron from the valence band, leaving a corresponding hole in the valence band; and we have already seen that such holes act as positive charge carriers. Silicon doped with boron is therefore known as a **p-type semiconductor**, indicating that it has extra (positive) holes in its structure that can work as charge carriers. A suitable energy-level diagram is given in Figure 8.12.

Most of the holes appear at the top of the valence band, with their concentration falling off nearer to the nucleus. In Figure 8.13, which shows the distribution of holes and electrons, the electrons have their highest concentration at the lower levels, and holes the opposite. This seems less strange when you realise that a hole needs *more* energy, not less, to get nearer the atomic nucleus. The concentration varies throughout the band because there are more electrons (or holes) that just have enough energy to cross the forbidden gap. Progressively fewer have sufficient extra energy to take them deeper into the band.

For the purposes of describing semiconductor operation, it is easier to use an energy-level diagram that refers specifically to the electrons and holes. We use minus signs (−) to indicate an area where electrons dominate, and plus signs (+) to show where there are more holes. Figure 8.13 looks the same as the energy-level diagrams we have used so far in this chapter, but we can use the shading to

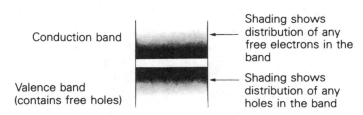

Figure 8.13 An energy-level diagram illustrating distribution of holes and electrons. This kind of diagram will be used extensively when discussing semiconductor devices.

show the probability of finding electrons or holes (as charge carriers) in the bands. You will discover how useful this diagram can be for describing the way semiconductor devices work, in the next chapter.

Before leaving Figure 8.13, let us be completely sure exactly what it represents:

1. It is an energy-level diagram for a piece of material as a whole.
2. The upper band is the conduction band. It may be empty or it may have electrons in it. Any electrons in the conduction band can move freely through the material, and each electron carries one unit of negative electric charge. The density of the shading corresponds to the probability of finding free electrons at any given level in the band. The space between the bands is the forbidden gap. No electrons can exist with this energy level, though they can cross the gap (apparently in zero time, although that's another one of the mysteries of physics . . .).
3. The lower band is the valence band. Its electrons are usually locked into the atomic structure, but in semiconductors there may be holes in it. Holes can drift about freely in the valence band, and are each carriers of one unit of positive electric charge. The density of the shading represents the probability of finding holes at any given level in the valence band.

We do not consider the absolute values of the energy levels, any more than we consider the inner bands of the atomic structure, for they are not relevant to the way semiconductor devices work. The energy is actually the total kinetic and potential energy of the electrons, and is a function of the physical structure of the material and also of any applied electrical potential. If an electric charge is applied to the material, the width and relative position of the energy bands will be unaltered, but the total energy in the system will change, moving the whole system of bands up and down the energy scale. Thus a negative electric potential applied to a material will move all the bands of the diagram up the energy scale, by adding to the energy (negative) of all the electrons.

An applied positive charge will move *all* the bands down the diagram. You will see this overall movement of energy bands in the next chapter, where we shall be looking at the simplest of common semiconductor devices, the *pn* diode.

Some of the theories and ideas that I have described in the first part of this chapter are not easy to understand at first. However, it is important

that you *do* understand them before going on to Chapters 9 and 10, because a lot of the explanations in those chapters are done through energy-level diagrams.

The manufacture of semiconductors

It is harder than you might think to manufacture a piece of *n*-type or *p*-type silicon. The first step is to purify the silicon, using the best chemical techniques available. 'Chemically pure' silicon may contain an almost negligible amount of impurity, but this is enough to make it quite useless for making semiconductor devices. Silicon in which the impurities are measured in parts per *billion* is needed. A technique known as zone refining was developed. Zone refining consists of taking an ingot of pure silicon, and repeatedly moving it through a radio-frequency heating coil in the same direction. Figure 8.14 illustrates this. The heated area sweeps all the impurities down to the end of the ingot, until eventually, after many passes through the coil, the major part of the bar is pure enough. The end of the ingot is cut off and sent back for chemical purification again.

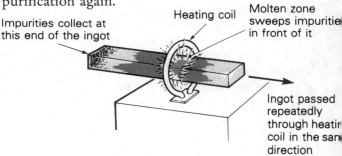

Figure 8.14 Zone refining a silicon ingot; the ingot is passed repeatedly through the heating coil, which sweeps all the impurities down to one end of the ingot.

Next, the silicon must be made into a crystal. Most solids, when they are cooled from a molten state, crystallise into many crystals, with distinct boundaries between each crystal. This is not good enough for semiconductor devices, for a single large crystal is needed. The silicon is heated to a fraction above its melting point in an inert container (that is, a container made of a material like quartz, which will not react with the silicon). A single tiny silicon crystal is dipped into the molten silicon, and then withdrawn very slowly; the molten silicon makes the crystal 'grow' and, if conditions are exactly right, a sausage-shaped single crystal, about 50 mm in diameter, can be drawn out.

The controlled impurities (called **dopants**) required to make p-type or n-type silicon must also be added. There are two possible ways in which this can be done. Most obviously, the required impurities can be mixed into the molten silicon before the crystal is drawn. This method is often used. A second method, less obvious but potentially much more useful, is to add the impurities by a process known as diffusion. If the crystal 'sausage' is cut into thin slices – known in the trade as **wafers** – the required dopants can be added to each wafer as required. The wafer is first heated to about 1200 °C (lower than the melting point of silicon) and then exposed to an atmosphere containing – for example – phosphorus. The phosphorus atoms diffuse into the silicon, and change it into n-type silicon. The usefulness of the process is that it is possible to control quite accurately the depth to which the phosphorus diffuses. This is important in making most modern semiconductor components.

The same process can be used to make p-type silicon, simply by using an atmosphere of boron instead of phosphorus.

It is even possible to add an insulating and protective layer of **silicon dioxide** by putting the wafer into an atmosphere of water vapour and oxygen at a high temperature. The diffusion process makes many things possible and is crucial to the manufacture of devices from diodes to integrated circuits, as we shall see in the next few chapters.

■ CHECK YOUR UNDERSTANDING

● It is useful to think of **electrons** as orbiting the **nucleus** of an atom in the same way as planets orbit the sun.

● The more energy an electron has, the further it will be from the nucleus.

● No electron may occupy exactly the same orbit as any other, and there are **forbidden gaps** at certain distances from the nucleus in which electrons cannot exist.

● Each permitted orbit – called a 'shell' – is a region consisting of lots of possible orbits, some more likely to be filled than others. Thus each shell can be regarded as a 'band' in which electrons can be found, with varying degrees of probability.

● We can use an **energy-level diagram** to represent a section through the various bands surrounding the nucleus of an atom. Because only the outer bands are involved in electronic interactions, energy-level diagrams are drawn without the inner bands.

● The outer shell – or band – in an atom is called the **valence** band. Beyond this is the **conduction** band, normally empty.

● Materials that are insulators have a big gap between the valence and conduction bands, which electrons cannot easily cross. Conductors have almost no gap between the conduction and valence bands.

● **Semiconductors** are materials that have a very small gap, and show characteristics halfway between those of conductors and insulators.

● **Intrinsic** semiconductors are semiconductors naturally.

● **Extrinsic** semiconductors are man-made by the addition of minute traces of elements to a material that is normally an insulator.

REVISION EXERCISES AND QUESTIONS

1 What are i) intrinsic and ii) extrinsic semiconductors?
2 What is i) a p-type semiconductor, and ii) an n-type semiconductor?
3 Which of the energy bands are important in electronic interactions?
4 In order for an electron to move to a higher energy band, does it need to gain or lose energy?

The *pn* junction diode

Introduction

The *pn* diode is the simplest semiconductor device that is of practical use, and the way it works is central to an understanding of more complex devices. This chapter looks at the diode in detail. There are a number of 'specialised' versions of the *pn* diode that make use of various properties of the *pn* junction, and we shall be looking at those, too.

The principles and physics of the *pn* diode

The special properties of semiconductors are used to make a wide variety of electronic components, ranging from very simple (diodes) to very complicated (microprocessors). A detailed understanding of one of the simplest semiconductor devices, the **pn junction diode**, is an important prelude to a study of more complicated devices, so it is here that we begin.

A **diode** is basically the electrical equivalent of a one-way valve. It normally allows electric current to flow through it in one direction only. The symbol for the diode is given in Figure 9.1. The arrowhead in the diagram shows the permitted direction of **conventional** current flow. (Remember that mistake (see Chapter 2)? Electron current goes the other way.) Diodes can be made

Anode   Cathode

Figure 9.1 The circuit symbol for a diode.

to pass currents varying from microamps to hundreds of amps; a selection of different kinds is shown in Figure 9.2.

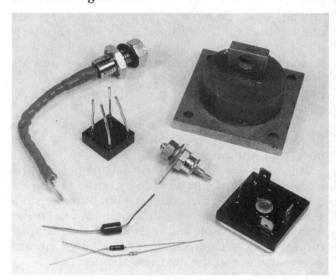

Figure 9.2 A selection of diodes of various power ratings, from 1 A to 20 A.

When a diode is passing current normally (as it does in one direction) it is said to be **forward-biased**. When it is preventing a current flow it is said to be **reverse-biased**. A 'perfect' diode would have infinite resistance in one direction, and zero resistance in the other. Since things are seldom perfect, a real diode exhibits rather different characteristics, the most notable of which is a **forward voltage drop**, which is constant regardless (almost) of the current being passed by the diode.

Compare Figures 9.3 and 9.4. In Figure 9.3 the diode is reverse-biased and does not conduct. The lamp remains out and the meter shows the whole battery voltage appearing across the diode. In

Figure 9.4 the diode is forward-biased, the lamp lights, and the meter is showing 0.7 V across the diode: the forward voltage drop. By using bulbs of different wattages to pass different currents through the circuit we can show that the 0.7 V is more or less constant.

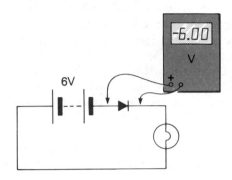

Figure 9.3 *A circuit to reverse-bias a diode. Note that the meter is connected with its '+' to the battery '−', which results in a negative reading on this digital meter.

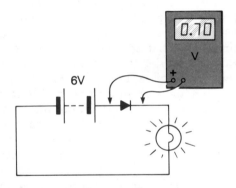

Figure 9.4 *A circuit to forward-bias a diode: the forward current is limited by the lamp.

The diode used in this demonstration is a **silicon** diode, and is typical of most of those in common use today. The forward voltage drop of a silicon *pn* diode is in the range 0.6–0.7 V. If we were to replace it with one using a **germanium** semiconductor, the forward voltage drop would be constant at between 0.2 and 0.3 V. The voltage drop therefore seems to be something to do with the semiconductor material itself. What causes it? To find out we must look at the way the diode functions at an atomic level, and go back to the energy-level diagrams.

Figure 9.5 shows an **energy-level diagram** for a piece of *p*-type silicon and a piece of *n*-type silicon; the two materials are not in contact. Now we place the two pieces of silicon in intimate contact

(fuse them together for best results) and the energy-level diagram looks like Figure 9.6.

A little thought will show what has happened. The *p*-type material has extra holes and the *n*-type material has extra electrons. When the two types of semiconductor are put in contact the 'extra' electrons in the *n*-type material flow across the junction to fill up the holes in the *p*-type material. The electrons drift from the *n*-type to the *p*-type through the conduction band, and then fall down into the valence band of the *p*-type material to annihilate one hole for each electron. (The word 'annihilate' is usually used when an electron occupies a hole and causes both of them to 'disappear'.) The holes also move, drifting from the *p*-type to the *n*-type semiconductor in the valence band.

> If you find difficulty in understanding the last paragraph, you should read Chapter 8 again.

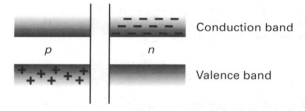

Figure 9.5 Energy levels for two separate pieces of semiconductor.

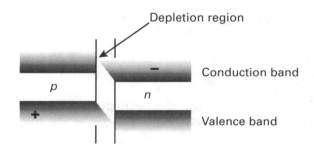

Figure 9.6 Energy levels for two pieces of semiconductor, one *p*-type and one *n*-type, in contact with each other.

Eventually an equilibrium is reached, but the *n*-type material has lost electrons to the *p*-type, and has lost energy in the process. The *p*-type has gained energy, and the result is the state of affairs represented by Figure 9.6.

The 'hill' in the diagram represents an area in the junction between the two semiconductor materials, typically about 0.5 μm wide, called the **depletion region**, or sometimes the **transition region**. In this region there are no holes or free electrons. The 'height' of the hill – the difference in energy levels – can be measured in volts: for silicon it is 0.7 V, and for germanium it is 0.3 V. Thus the question about the diode's forward voltage drop is answered.

The junction between the *p*-type and the *n*-type semiconductor is, in fact, the **active** part of the *pn* diode. A typical low-power *pn* junction diode might be constructed as shown in Figure 9.7.

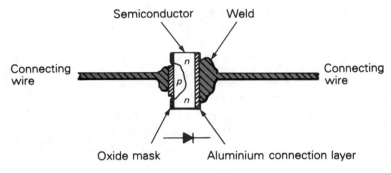

Figure 9.7 The construction of a low-power diode. The circuit symbol underneath indicates the direction of current flow.

Operating conditions for the *pn* diode

The *pn* diode with reverse bias

If a *pn* diode is connected as in Figure 9.3 (reverse-biased), the energy-level diagram for this state will be as shown in Figure 9.8.

The energy of the *p*-type material is increased by the negative battery potential, while that of

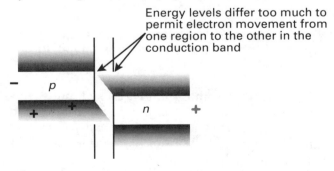

Figure 9.8 Energy levels in the reverse-biased diode.

the *n*-type is decreased. Look at the thickness of the conduction and valence bands and at their relative positions in the two different materials, on either side of the depletion region. No parts of the conduction bands or the valence bands have the same energy levels. Electrons cannot move from the *p*-type to the *n*-type material through the conduction band because even the least energetic electrons have too much energy. Similarly, holes in the valence band of the *n*-type material all have too little energy to drift into the valence band of the *p*-type. The energy levels of the bands on either side of the depletion region are too different to allow either electrons or holes to move from one type of material to the other.

It follows that if electrons and holes are unable to cross the depletion region, there can be no current flow.

The *pn* diode with forward bias

Let us reverse the situation and connect the diode as in Figure 9.4, to forward-bias it. The energy-level diagram for this state is given in Figure 9.9. This time, the battery potential decreases the energy of the *p*-type semiconductor and increases the energy of the *n*-type. When the potential applied is equal to or more than 0.7 V (assuming that we are using a silicon diode and not a germanium one) the two types of material are at the same energy level, and the conduction and valence bands coincide. Both electrons and holes can move freely between the two types of semiconductor, and electric current can flow through the diode.

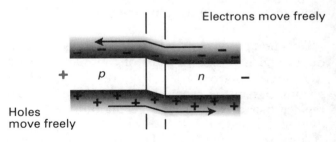

Figure 9.9 Energy levels in the forward-biased diode.

The reason for the 0.7 V forward voltage drop of the silicon diode should now be even clearer. A potential of 0.7 V must be applied to the *pn* junction to equalise the energy levels, and this is effectively subtracted from the potential available to push electric current through the diode.

Make sure you understand these explanations. They are vital to understanding the principles of transistors, described in the next two chapters.

Power dissipation in a *pn* diode

It is the forward voltage drop that determines the upper limit of the current that can safely be passed through a given diode. Let's see why.

Here is an example. The lamp in Figure 9.4 might use a current of 1 A, and the battery might be 12 V. In such a circuit, a current of about 1 A flows in the circuit, but some of the power is 'wasted' in the diode as a result of the forward voltage drop, *and is dissipated from the junction as heat*.

The amount of power that the diode dissipates can be calculated very simply using the formula

$$P = V_f I$$

where P is the power (in watts), I is the current flowing (in amps), and V_f is the forward voltage drop of the diode in question. With 1 A flowing, the diode will dissipate 0.7×1 watts, or 0.7 W. Any diode used in a circuit must have a **power rating** that exceeds the expected dissipation, preferably by a comfortable margin.

In practice, the formula is complicated by the fact that the diode gets hot. The intrinsic conductivity of semiconductors increases with temperature (the bands get thicker) and this has the effect of slightly lowering the forward voltage drop as the temperature rises. For both silicon and germanium the forward voltage drop decreases at the rate of about 2.5 mV/°C temperature rise.

The junction temperature of a silicon diode may exceed 100 °C with the diode operating near its maximum current, so temperature effects can become quite significant. The 'normal' forward voltage drop of 0.7 V would be reduced to 0.45 V for a 100 °C temperature rise. For most purposes this effect can be discounted, but in circuits where the voltage drop across the diode is important, we have to bear it in mind.

Silicon is much more tolerant of temperature than germanium. For this reason, and also because the **reverse leakage current** (see below) is much lower, germanium diodes are rare; they are used only in a very few applications where a low forward voltage drop is essential.

Reverse leakage current in a *pn* diode

When connected in a circuit that reverse-biases it (Figure 9.3 again) a diode doesn't block the current completely, but allows a very small current to flow. This current is called the **reverse leakage current** and is partly due to thermally generated electron–hole pairs and partly to leakage across the surface of the diode. The reverse leakage current is very small for silicon diodes, in the region of 2–20 nA. For almost all purposes this minute current can be ignored. The reverse leakage current for germanium diodes is much higher, typically 2–20 µA, and this may be important in some circuits.

The reverse leakage current is temperature-dependent; for both silicon and germanium it roughly doubles for each 10 °C rise in temperature.

Reverse breakdown in a *pn* diode

As the voltage applied to a diode to reverse-bias it is increased, there comes a point at which **reverse breakdown** occurs. This breakdown is permanent and destroys the diode. Depending on the construction of the diode, the point at which reverse breakdown occurs can be anywhere from 50 to a few hundred volts. Manufacturers always specify the safe reverse-bias voltage, which is normally referred to as **peak inverse voltage**. Thus a 50 p.i.V. diode can be safely reverse-biased at up to 50 V.

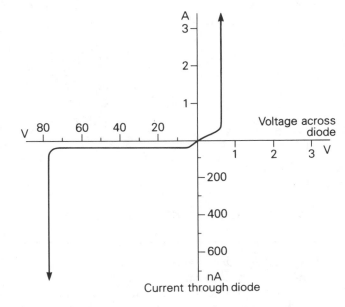

Figure 9.10 The conduction characteristics of a silicon *pn* junction diode.

We can draw a graph of a typical *pn* diode's conduction characteristics at room temperature (Figure 9.10). Note that the four scales of the graph are different, to show different aspects of the characteristics. On the right, the voltage across the diode is shown for different values of current through the diode. The graph continues straight up until the current reaches a value that causes overheating ($P = V_f I$) and melts the *pn* junction. On the left, the reverse leakage current is shown for increasing voltage. It remains steady at a few nanoamps until reverse breakdown occurs (at a substantially higher voltage than the 50 p.i.V. manufacturer's maximum for this device) and the current increases to a value limited by factors other than the characteristics of the *pn* junction, which at this point will have just ceased to exist.

Special types of diode

The Zener diode

The Zener diode makes use of a form of reverse breakdown to provide a constant 'reference' voltage, as required by many circuits. The circuit of Figure 9.11 shows the symbol for a Zener diode and also a suitable circuit to demonstrate its properties. In this circuit the diode is reverse-biased (Zener diodes are always used this way) and the meter is showing a drop of 2.8 V across it.

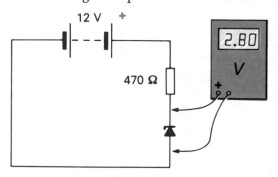

Figure 9.11 ★A Zener diode in a typical circuit.

> The voltage drop across a reverse-biased Zener diode will be substantially constant for all values of current up to the diode's limit, and for all values of p.d. higher than the Zener voltage.

A resistor is necessary to limit the current through the circuit. In this case the current is calculated by Ohm's law:

$$\frac{12 - 2.8}{470} = 19.6 \text{ mA}$$

If we were to double the battery voltage to 24 V, the current through the circuit would increase to

$$\frac{24 - 2.8}{470} = 45 \text{ mA}$$

but the voltage across the Zener diode still remains constant at 2.8 V. The Zener diode is therefore a valuable device for voltage regulation, giving a fixed reference voltage for a range of input voltages. According to the doping and the construction used, Zener diodes can be made with any of a range of forward voltage drops.

Zener diodes are named after C.M. Zener, who in 1934 described the breakdown mechanism involved. In fact, Zener's description applies only to diodes with a Zener voltage of less than about 3 V. Above this, a mechanism called **avalanche breakdown** begins to take over. However, both types are generally lumped together under the heading 'Zener diodes', and are available in a range of voltages from 2 to 70, and with power ratings from 500 mW to 5 W. The power dissipation of the Zener diode is calculated by the usual formula

$$P = V_Z I$$

where V_Z is the Zener voltage. By way of example, the Zener diode in Figure 9.11 is dissipating

2.8 × 19.6 = 54.9 mW

The conduction characteristics of a Zener diode can be drawn in a diagram similar to Figure 9.10. Look at Figure 9.10 again: the only difference in the Zener diode's characteristics is in the reverse breakdown (lower left quadrant), which will occur at the Zener voltage. The important factor – not revealed by the diagram of characteristics – is that a Zener diode is not damaged by reverse breakdown, provided that the current through the diode is limited to a safe value.

The varicap diode

The junction of a reverse-biased diode has a measurable **capacitance**. The *p*-type and *n*-type regions form the **plates** of the capacitor, and the depletion region acts as a **dielectric**. You will remember that the capacitance of a capacitor depends (among other things) on the thickness of the dielectric, and it is this fact that makes the *pn* diode useful as a **variable capacitor** with no moving parts.

The diode is reverse-biased. If the biasing voltage is then increased, the difference in the energy levels of the *p* and *n* regions is made greater. As this happens, a greater thickness of the junction is depleted of charge carriers, increasing the width of the depletion regions. Effectively, the capacitor's dielectric has become thicker, and this lowers its capacitance.

A **varicap diode** (a composite word from '*vari*able *cap*acitance') is a diode in which the change of capacitance with applied reverse voltage is enhanced as far as possible by the diode's physical construction. The maximum capacitance and the amount of change are both small, although they are sufficient to enable the varicap diode to be used in the tuning circuits of a television, for example.

Photodiodes

A photodiode is structurally very similar to a normal *pn* junction diode, though there may be a mechanical difference brought about by the necessity to maximise the area of the junction that can be exposed to light. Photodiodes are used in **reverse-biased mode**, and the leakage current will then depend on the amount of light falling on the device. Photodiodes are useful for measurement applications, since the leakage current is directly proportional to the light intensity over a wide range.

Silicon is generally used for photodiodes, so the peak response to light is in the infrared region. The actual amount of current is also rather small. A typical photodiode might have a dark current of 1.5 nA, and an output current in bright sunshine of 3.5 μA. Substantial amplification is therefore required for most applications.

Light-emitting diodes

A **light-emitting diode** (LED) is an extremely useful device, and can be used instead of a miniature incandescent lamp in a whole range of applications. LEDs are usually ruby red, but green and yellow versions are available, although they are a little more expensive and therefore less common.

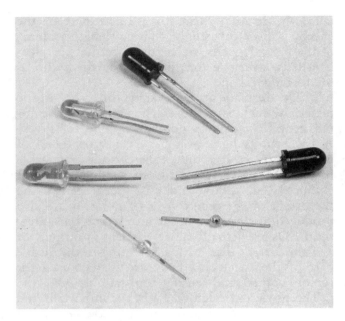

Figure 9.12 Light-emitting diodes.

Figure 9.12 shows typical LEDs.

Like ordinary *pn* diodes, LEDs have no inherent current limiting characteristics, and must be used with a resistor to limit the current flowing through them, generally to around 20 mA for a small LED indicator. A suitable circuit is shown in Figure 9.13. LEDs are used in the forward-biased mode: indeed, they must be protected from reverse bias as they usually have a reverse breakdown voltage of only a few volts. If breakdown occurs, the LED is destroyed.

Mechanism of light emission

To move an electron from the valence band into the conduction band requires energy; an example of this is the creation of an electron–hole pair. Similarly, if an electron falls down from the conduction band to annihilate a hole in the valence band, energy is emitted. In *pn* diodes this energy

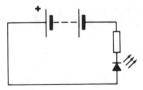

Figure 9.13 A typical circuit for driving a LED. The resistor is essential to limit the current through the LED.

takes the form of heat. In a LED, part of the energy is given off as light; the wavelength (colour) of the light depends on the distance that the electron falls, which in turn depends on the width of the forbidden band.

LEDs are made from a variety of rather exotic materials, such as gallium arsenide phosphide for red or yellow, and gallium phosphide for green. A LED emits light over a very narrow band of wavelengths, so it is unfortunately not practicable to use filters over the LED to produce different colours. Red, yellow and lime-green seem to be the extent of the repertoire so far. LEDs are also produced to emit light in the infrared region. Infrared LEDs are actually more efficient in output for a given current than LEDs operating in the visible part of the spectrum. They are widely used for remote control and sensing applications.

The physical design of a LED is important and difficult. The semiconductor material is rather opaque, so most of the light produced never reaches the surface. It's a little like having an electric torch immersed in a bucket of tar! To make the best of it, the junction has to be very close to the surface, only about 1 μm or so, so that some of the light can escape. The p or the n region therefore has to be very thin, and electrical connections to this layer have to be made in such a way that they do not obstruct the light. Figure 9.14 shows the construction of a typical LED.

The LED has no time lag on the production of light (such as there is on a normal tungsten filament lamp) and so can be switched on and off very rapidly. In practical applications LEDs are often **pulsed**. This is because the light output increases quite rapidly with increasing current, so a LED pulsed at 50 mA, so that it is on half the time, will look brighter than it would with a continuous 25 mA, while dissipating the same amount of heat. The pulse-repetition frequency has to be rapid enough for persistence of vision to make it look as if the LED is on continuously: more than about 50 Hz.

Like a pn diode, the LED exhibits a forward voltage drop. This varies according to the type and colour of the LED but is generally between 2 and 3 V.

LED displays

LEDs are often used in multiple displays; a LED ten-bar array is shown in Figure 9.15. The bar array is designed so that it can be stacked with other identical arrays, to make a bar graph display of any length required. Such indicators are often used for record level meters on hi-fi systems, as they are cheaper, more accurate, more robust and have more 'sales appeal' than ordinary mechanical meters.

LEDs are also used for the familiar seven-segment display, of the kind commonly used in digital alarm clocks. The seven-segment display can be used to produce any of the digits 0–9, and a very limited number of letters. The basic form of the display, together with a set of digits and letters, is given in Figure 9.16.

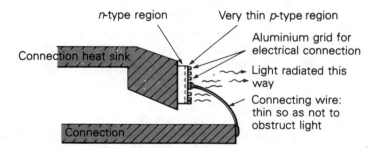

Figure 9.14 The physical construction of a LED. Light is emitted through the very thin p-type region and the device must be designed so that this light is obstructed as little as possible. A large connection area is made to the n-type region, and this also serves to conduct heat away from the junction.

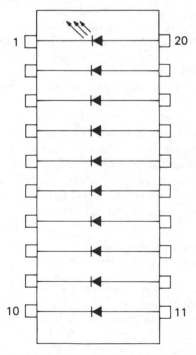

Figure 9.15 A ten-LED bar graph display.

Figure 9.16 A basic seven-segment display, along with some of the characters that can be produced.

Figure 9.17 shows some typical modern LED displays.

Multiplexing displays

A typical alarm-clock display, equipped with hours, minutes and seconds, would require six seven-segment displays. If they were all connected, with seven connections to the bars and a common anode or cathode, there would be eight connections to each digit, or 48 connecting wires altogether. This is not very economical, and is wasteful of power, when you consider that at 12:58:58 there will be no less than 30 segments illuminated, each taking about 15 mA, a total consumption approaching 0.5A, and a power dissipation, in the LEDs alone, of nearly 1 W.

In practice the display is **multiplexed**, and is connected as shown in Figure 9.18. This cuts the

number of wires down to 13, and adding extra numbers requires only one more wire for each extra digit. **Multiplexing** (or sometimes **time-division multiplexing**) is a technique in which a circuit or system divides a task (or signal) up into 'time-slices' instead of trying to do everything at once.

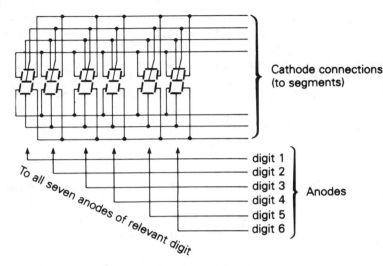

Figure 9.18 Multiplexing a six-digit seven-segment display.

Circuits for driving multiplexed LED displays are specially designed to make use of the display economically, and in practice only one digit is illuminated at any one time. It works like this. The cathodes of the relevant segments for the first digit are switched to the power supply, and the digit-1 anode is made positive for a short time. The digit-1 anode is then disconnected and the cathodes switched in a switch pattern for the second digit. The digit-2 anode is now made positive for a short time, and the sequence repeated over and over for each digit in turn.

Only one digit is ever on at any one time, but if the sequence is repeated at a high-enough speed, the human eye's persistence of vision will make it look as if the six digits are all lit up together. The multiplexing frequency is usually 1 kHz or so.

It might seem as if the circuits required for multiplexing the display would be very complicated; this is indeed the case, but they can be made very cheaply as an **integrated circuit**.

Figure 9.17 LED displays.

■ CHECK YOUR UNDERSTANDING

● A **diode** is a component constructed from a piece of *n*-type material in contact with a piece of *p*-type material. It will pass an electric current in one direction, but not the other.

● Diodes always have a **forward voltage drop**, which is independent of the current flowing through them. For silicon it is about 0.6–0.7 V, and for germanium it is about 0.2–0.3 V.

● The amount of current a diode can carry depends on its ability to dissipate waste heat.

● Waste heat is generated in a diode because of the forward voltage drop.

● **Zener** diodes are used in reverse breakdown mode. The voltage across them – up to a few tens of volts – is more or less constant for any higher applied voltage.

● **Varicap** diodes change capacitance with applied voltage.

● **Photodiodes** have a reverse leakage current that increases with the amount of light falling on the junction.

● **Light-emitting diodes** produce visible or infrared light when a current is passed through them.

1 Draw a circuit diagram for a *pn* diode, labelling the anode and cathode and indicating the direction of conventional current flow.

2 A silicon diode is passing a current of 2 A without getting noticeably hot. Roughly how much power is it dissipating?

3 Draw a circuit for illuminating a red LED, using a 9 V battery.

4 Draw a single-digit seven segment display. Shade in the segments that are illuminated to make a figure '3'.

5 What are i) photodiodes and ii) varicap diodes?

6 What would this circuit do: i) with a 6 V power supply connected positive to T_1, negative to T_2; ii) the same power supply connected the other way round, with positive to T_2; iii) a 3 V power supply connected as in i)?

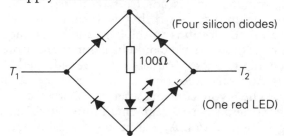

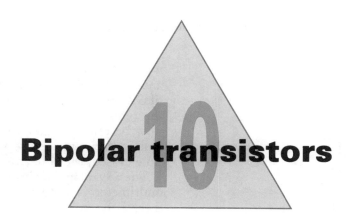

Bipolar transistors

Introduction

The transistor is the most fundamental device in modern electronics. The invention of the transistor began the 'electronics revolution', and the transistor is still the most basic of all elements in an electronic circuit.

There are two main classes of transistor: **bipolar transistors** and **field-effect transistors**. The first group is the subject of this chapter.

Description

A bipolar transistor has three terminals, referred to as the **emitter, collector** and **base**. The bipolar transistor is a **current-operated device**. It amplifies a signal by providing an output current that is proportional to the input current multiplied by a factor known as the **gain**.

Because transistors are constructed with three layers of semiconductor in a sort of 'sandwich', it is possible to make two different types, known as *npn* and *pnp*. The letters represent the type of semiconductor material used, either *p*-type or *n*-type. Both kinds of transistor are manufactured (although *npn* is probably more common) and are used in identical circuits *except* that the supply voltages are opposite: *npn* transistors are connected with the emitter to the negative supply, whereas *pnp* transistors are connected with the emitter towards the positive.

The circuit symbols for the two types of bipolar transistor are given in Figure 10.1. Note the direction of the arrowheads in the diagrams. Like similar arrows in other component symbols, they show

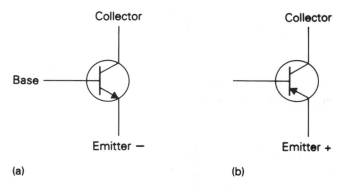

Figure 10.1 Circuit symbols for (a) *npn* and (b) *pnp* transistors.

the direction of **conventional current**, not electron flow.

The bipolar transistor behaves as if it were a variable resistor, the value of which depends on the current flowing through the base connection. The name 'transistor' is a contraction of 'transfer resistor', which goes some way to describing the properties of the device.

The circuit in Figure 10.2(a) shows a transistor in a test circuit (but don't try it!), connected in what is known as **common emitter mode**. If the input terminal A is made positive with respect to the emitter, a current will flow between the base and emitter. The base–emitter junction will behave in exactly the same way as a forward-biased semiconductor diode. Within a certain range of base currents, the collector–emitter junction will exhibit the characteristics of a variable resistor, the resistance of which is inversely proportional to the base current (Figure 10.3). The output current of the transistor, measured on a meter in series with the collector, will be larger than the input current by a fairly large constant: typically 20–1000, depending on the specific device.

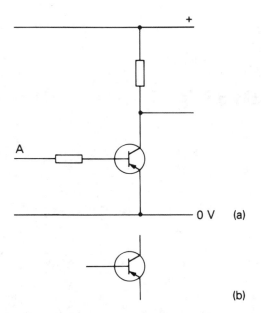

Figure 10.2 A simple transistor amplifier (with an indication of the base circuit).

Figure 10.3 The transistor places a variable amount of resistance in the collector circuit.

A practical circuit for measuring a transistor's output current for a range of input currents is given in Figure 10.4. Adjustment of the variable resistor VR_1 permits control of the base current, and makes it possible to check the output current for a range of input currents. The circuit is used as follows;

with SW_1 open, and VR_1 set to a maximum resistance, the meter is connected across the terminals of SW_1 to read the base current. VR_1 is set to give the required value of base current. The meter can now be removed and connected in series with the collector circuit, across terminals A–A. If SW_1 is now closed, the collector current can be measured.

The advantage of this test circuit is that only one multimeter is required. The disadvantage is that there is no compensation for the resistance of the meter, which may affect the accuracy of the experiment. If required, a resistor equal in value to the resistance of the multimeter at the current range selected may be wired in series with SW_1 for more accurate results.

Using the components specified in Figure 10.4, the test can be performed for values of collector current up to about 200 mA. The battery may begin to flag at currents in excess of this, and if the collector circuit is left connected for any length of time, the transistor will get quite warm: after all, at 200 mA (and, once again, assuming that the test meter has negligible resistance) it will be dissipating

$$P = VI \text{ watts}$$
$$P = \frac{9 \times 200}{1000}$$
$$P = 1.8 \text{ watts}$$

For this reason, the specified transistor (an extremely high-power type) should be used in this circuit. Resistor R limits the base current to a safe value.

Over the range of collector currents from 10 to 200 mA, the calculation I_c/I_b (collector current/base current) should yield a value that is approximately constant. This constant is the **large-signal current gain** of the transistor when used in the common emitter mode. The large-signal current

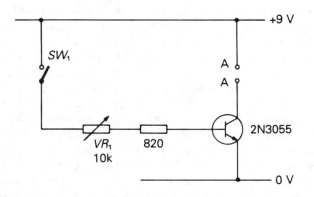

Figure 10.4 *A practical circuit for approximate measurement of transistor gain.

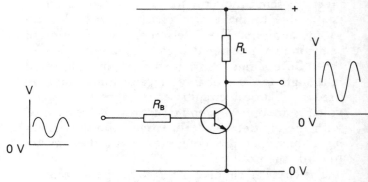

Figure 10.5 A simplified transistor amplifier, unlikely to be used in practice.

gain is referred to by the symbol h_{FE}. That of the 2N3055 is quoted (by a manufacturer) as being between 20 and 70; you will find quite a wide variation between different examples of the same device.

It should now be clear how the transistor works as an amplifier. Figure 10.5 shows a transistor amplifier circuit at its simplest. A voltage applied to the input of this amplifier will cause a base current to flow, and the base current will be reflected in a change in collector current, which in turn will alter the voltage appearing at the output. A convenient way of visualising this is to think of the transistor's collector circuit and the collector load resistor R_L as forming two halves of a potentiometer, with the lower half variable in value as the base current changes. Figure 10.6 illustrates this.

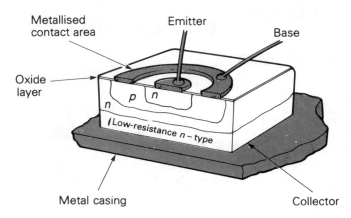
Figure 10.7 The physical construction of a silicon *npn* transistor.

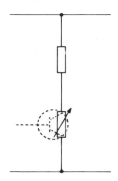
Figure 10.6 The amplifier's output circuit as a potential divider.

Physics

Having briefly examined a typical bipolar transistor, we now turn to the physics of the device and to the way it has to be constructed. A typical transistor is shown schematically in Figure 10.7. It is a three-layer device, not dissimilar to a *pn* diode in general appearance except that it has an extra layer; compare Figure 10.7 with Figure 9.7 in Chapter 9. An important piece of information – you will see why it is important as the description of the way the transistor operates progresses – is that the *n*-type material is more heavily doped than the *p*-type material. The result is that the number of free electrons in the *n*-type semiconductor greatly exceeds the number of holes in the *p*-type semiconductor.

We can draw an energy-level diagram for the transistor, first when the device is 'at rest' with no power supply connected to any of the terminals. Figure 10.8 shows this. Neither electrons nor holes can cross the depletion regions, because of the large difference in energy levels. If we now make the collector positive with respect to the emitter, as in the circuit shown in Figure 10.5, the picture is different, changing to that shown in Figure 10.9. Clearly, however, there is still no flow of electrons or holes through either junction.

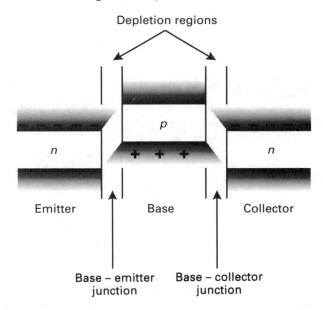
Figure 10.8 Energy-level diagram for the transistor with no voltage applied.

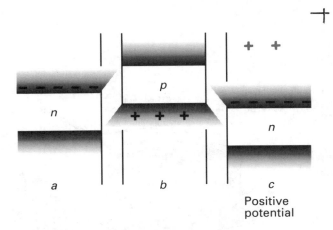

Figure 10.9 Energy-level diagram when the collector is positive relative to the emitter.

When the **base** is made rather more positive than the emitter, the state of affairs changes dramatically, as illustrated by Figure 10.10, in which a positive voltage is applied to the base to make it positive with respect to the emitter (but still negative in relation to the collector). Under these conditions, large numbers of electrons are able to flow from the emitter base region, the energy levels at the emitter–base junction having been made sufficiently similar to allow electrons to move across the junction. A few of these electrons combine with holes in the base of the region. Restoration of the holes in the *p*-type semiconductor requires a flow of electrons out of the base region (or a flow of holes into the base region). This constitutes the **base current** through the circuit.

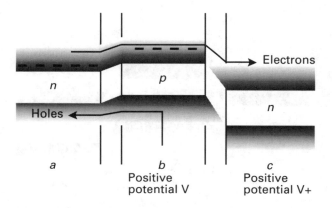

Figure 10.10 Energy-level diagram when the base is more positive than the emitter.

Most of the electrons in the base region drift towards the base–collector junction, where the positive potential attracts them across the depletion region into the *n*-type material of the collector.

Electrons can thus flow from the emitter to the collector, and this current flow constitutes the **collector current**. The collector current is larger than the base current because the relative scarcity of holes in the lightly doped *p*-type base region limits the amount of current that can be drawn from the base by restricting the availability of charge carriers.

Because both electrons and holes are involved in its operation, this type of transistor is called **bipolar**; and because of its construction, with a *p*-type base region and *n*-type emitter and collector, it is called an *npn* **transistor**.

It is equally possible to make a *pnp* **transistor**, having an *n*-type base and *p*-type emitter and collector. Figure 10.11 shows the circuit symbol for a *pnp* transistor, in a circuit similar to the one in Figure 10.5.

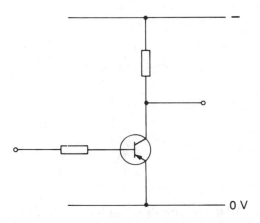

Figure 10.11 A *pnp* transistor used in a simple amplifier circuit.

> Note that for the *pnp* transistor the polarity of the power supply is also reversed, compared with the *npn* transistor.

The base and the collector are at a negative polarity relative to the emitter. The physics of its operation are similar to those of the *npn* transistor, but with the relative functions of holes and electrons exchanged.

Both *npn* and *pnp* transistors are readily available.

Almost all modern transistors are made using silicon as the semiconducting material. Compared with germanium, silicon is less affected by temperature variations, and has fewer undesirable characteristics.

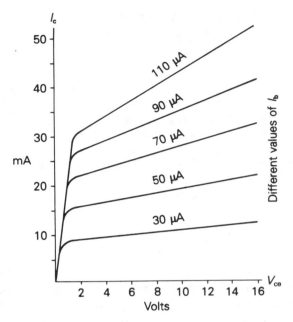

Figure 10.12 Characteristic curves for a transistor.

Characteristics

The circuit configuration shown in Figure 10.5 and 10.11 is referred to as a **common-emitter** circuit, the emitter being common to both input and output circuits. It is the most usual way of connecting a transistor. When used in this (or indeed, any other) mode, the transistor exhibits certain characteristics that we can use as a measure of its performance.

First, and perhaps most important, is the **large-signal current gain**, the value we measured at the beginning of this chapter, expressed as the ratio I_c/I_b, collector current divided by base current, and given the symbol h_{FE}. Unfortunately for those of us who like life to be simple, h_{FE} is not constant, but varies with collector voltage. This is known as the **Early effect** (after the man who first suggested what caused it). One of the best ways of illustrating the way that h_{FE} changes with collector voltage is to plot a graph of collector current (I_c) against

Figure 10.13 A selection of heat radiators and heat sinks.

collector voltage (V_{ce}) for a range of different base currents (I_b), choosing values for I_b and V_{ce} that are sensible for the transistor being measured. A typical set of graphs – called **characteristic curves** – is shown in Figure 10.12, which is worth careful study.

The straight, sloping part of the curve is called the **linear region**, and the transistor should be operated on this part of the curve for most purposes: for example, for all types of amplifier. The almost-vertical part of the curve marks the **saturated region** of the characteristic, where the transistor 'switches' rapidly from a non-conducting state (from 'off' to 'hard on') as the base current is increased. Transistors are operated in this part of the characteristic in digital circuits, where rapid switching and low power dissipation are important.

In the context of *pn* diodes, I mentioned **leakage current**. Transistors also have a leakage current, which flows into the collector when the base connection is disconnected (that is, open-circuited). The base leakage current is given the symbol I_{ceo}.

For silicon transistors I_{ceo} is very small, seldom as much as a few microamps. Germanium transistors have a much larger value of I_{ceo}, often approaching a milliamp. In germanium transistors the leakage current is also intensely temperature-dependent: so much so that they can be used as temperature-sensing devices. Silicon transistor leakage current is less affected by temperature at normal operating temperatures.

Temperature changes also affect the voltage drop at the base–emitter junction. For each 10 °C rise in temperature the base–emitter voltage is reduced by about 20 mV. Increasing temperature also changes the characteristic curves, spreading them further apart and moving them upwards on the graph. Finally, an increase in ambient temperature reduces the power-handling ability of a transistor by slowing the rate of heat transfer away from the casing (and thus the junction). Heat transfer from the transistor junction to the outside world is proportional to the temperature difference between them. A transistor's ability to dissipate heat is quoted in °C/W, being the temperature rise of the case (not the junction) per watt of dissipation at given ambient temperature, usually 20 °C. Metal **heat radiators** or **heat sinks** can be used to improve the heat-transfer capability of a transistor; Figure 10.13 shows some typical forms.

How to specify bipolar transistors

There are many thousands of different types of transistor, with every possible different specification. Manufacturers are continually introducing new designs and withdrawing obsolete types. There are what are known as 'design' types (the newest, for new equipment), 'stock' types and 'service' (obsolete) types. When a designer specifies a particular kind of transistor, the following main parameters are taken into account.

Class

There are at least three classes of semiconducting device according to reliability, capacity for withstanding unpleasant environments – such as heat, damp and radiation – and physical construction. One large semiconductor manufacturer calls these classes 'military', 'industrial' and 'entertainment'. The best – military – components are substantially more expensive than the other two classes.

Power

There are big transistors and small transistors. The big ones can dissipate as much as 150 W, whereas the smallest may overheat above 100 mW. Transistors with large dissipations are called **power transistors** and are often designed so that they can be bolted to a suitable heat radiator. The dissipation is partly a function of the size of the semiconductor chip, and partly a function of the arrangements inside the transistor for conducting heat away from the junction. A power transistor and a small plastic-encapsulated transistor are compared in Figure 10.14. Power dissipation is given the symbol P_{TOT}.

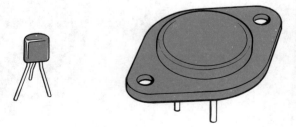

Figure 10.14 Large and small transistors compared.

Current

The maximum continuous collector current that a transistor can pass is given the symbol $I_{C(MAX)}$. It is related to power, in that power transistors usually have a larger value of $I_{C(MAX)}$ than small design transistors, but the power is of course a factor of current *and* voltage. Most transistors can withstand collector currents several times larger than their $I_{C(MAX)}$ if the currents are of very short duration.

Voltage

A useful measure of the working voltage of a transistor is the maximum safe voltage that can be applied between the collector and the emitter when the base connection is open-circuited. This is given the symbol V_{ceo}, and may be anywhere between ten and a few hundred volts.

Gain

We have already dealt with h_{FE}, the most convenient measure of a transistor's gain. There are other ways of measuring the gain (in common-base mode, for instance), but h_{FE} is the most commonly quoted and used parameter. The gain may be measured for a small signal instead of a large signal, in which case the symbol h_{fe} is used. More than any of the other parameters, the gain may vary from one example to another of the same type of transistor: a variation of ± 50 per cent is quite usual.

Frequency

A transistor's ability to 'follow' high-frequency signals depends on many factors, but principally on the width of the base region: the narrower the base region, the quicker electrons can cross it. Small signal transistors (*npn* silicon planar) would typically have a maximum usable frequency limit – this is given the symbol f_T – of 50–300 MHz. Power transistors have much lower frequency limits: up to 100 MHz is typical for a 10 W type. At present the maximum f_T available, outside very expensive specialist devices, is around 1 GHz, but the power would be very low.

Case

There are many different case designs, but most transistors use one of a relatively small number of shapes that have become 'standard' over the past few years. For small signal transistors, the **TO18** metal encapsulation or the **TO92** plastic (cheaper) are commonly used. In these, as in all transistor case designs, the shape of the case and the position of the leads indicate which of the three wires is which; they are not usually marked 'c', 'b' and 'e'. For transistors having a P_{TOT} of more than 500 mW or so, a larger metal case is used, generally the **TO5** design. For power transistors, a plastic **TO126** or **TO202** encapsulation with a metal tag for bolting to a heat sink or heat radiator might be used; and for the larger power transistors (5–150 W) the **TO3** case is usual; this one is unequivocally intended to be used bolted to a heat sink. All six of these common types are shown in Figure 10.15.

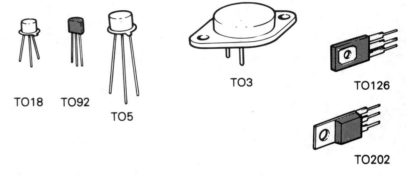

Figure 10.15 Six common types of transistor encapsulation.

■ CHECK YOUR UNDERSTANDING

● A **bipolar transistor** is a three-terminal amplifying device.
● The most usual circuit employs the transistor in **common-emitter mode**.
● A small current flowing between the transistor's **base** and **emitter** allows a much larger current to flow through the third terminal, the **collector**.
● The amount by which the collector current of a transistor can exceed the base current is called the **gain** of the transistor.

● Transistors are specified by: class (military/industrial/entertainment), power rating (in W or mW), current (in A or mA), voltage (from about 10 V to 500 V), gain (from about 10 to 1000), frequency (from about 100 kHz to 1 GHz), and case type (plastic, glass or metal; many different designs).

● Transistors are manufactured as *npn* or *pnp* devices. The two kinds are used in identical circuits, except that the power supply is reversed: in common-emitter mode *npn* transistors have the emitter connected to the negative supply, whereas *pnp* transistors have the emitter connected to the positive supply.

REVISION EXERCISES AND QUESTIONS

1 Draw the circuit symbols for *npn* and *pnp* transistors, indicating the emitter, base and collector terminals of each.
2 Explain the following expression:
   i)   large signal current gain;
   ii) heat sink;
   iii) leakage current.
3 Describe in one sentence the basic operation of a transistor.
4 Explain what is meant by the 'emitter current' of a transistor in a circuit.

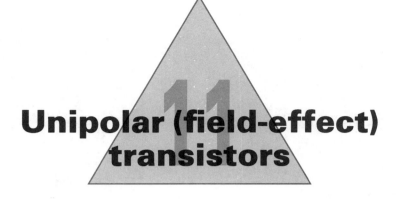

Unipolar (field-effect) transistors

Introduction

Field-effect transistors are a more recent development than bipolar transistors, and make use of a completely different mechanism to achieve amplification of a signal. Field-effect transistors (FETs) are **unipolar**, and only one type of charge carrier (electrons or holes) is involved in their operation.

There are two major classes of FET: **junction-gate field-effect transistors** (JUGFETs) and **insulated-gate field-effect transistors** (IGFETs). There are also, as we shall see, subdivisions within these two groups.

Junction-gate field-effect transistors

The main distinguishing characteristic of FETs, compared with bipolar transistors, is the fact that they are **voltage-controlled** rather than current-controlled. The circuit symbol is given in Figure 11.1. A **voltage** applied to the gate (and not a gate-emitter **current**) is varied to provide a corresponding change in resistance between the **source** and the **drain**. Unlike the bipolar transistor's base con-nection, the **gate** of the FET has a very high input resistance, at least a few tens of megohms and, in some cases, gigohms. The amount of current drawn by the gate is therefore extremely (sometimes unmeasurably) small.

FETs can be used in amplifier circuits, just like bipolar transistors. Compare the circuit of Figure 11.2, which shows a typical JUGFET, with that given in Figure 10.5 in Chapter 10. An obvious difference is the lack of a gate (base) resistor; because negligible current flows in the gate connection, such a resistor would make no difference to the operation of the circuit, merely adding a moderate amount of resistance to one that is huge in the first place.

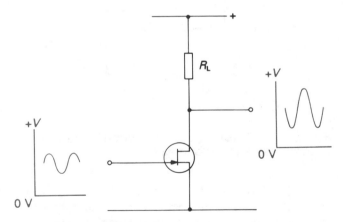

Figure 11.2 A simplified amplifier using an *n*-channel FET.

Just as there are *npn* and *pnp* transistors, so there are ***n*-channel** and ***p*-channel** JUGFETs. Figure 11.3 shows a *p*-channel version, while those in Figures 11.1 and 11.2 are *n*-channel (note the arrowhead).

Figure 11.1 Circuit symbol for an *n*-channel field-effect transistor.

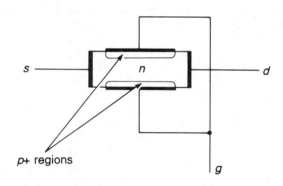

Figure 11.3 A simple amplifier using a *p*-channel FET (note the different polarity of the power supply, compared with Figure 11.2).

Junction-gate field-effect transistor physics

The JUGFET has a physical structure that can be represented by a diagram like the one in Figure 11.4. (In practice it is not easy to diffuse impurities with both sides of the wafer, and a rather different layout is used.)

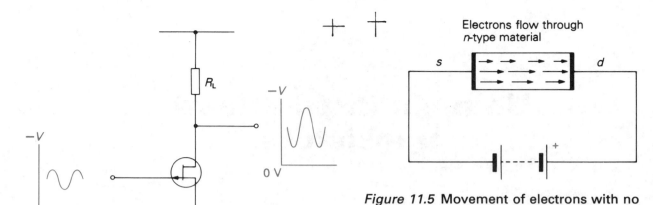

Figure 11.5 Movement of electrons with no voltage applied to the gate.

Now observe the effect of a negative potential applied to the gate regions. The junction between the *p* and *n* regions forms a reverse-biased diode so no current flows, but an electric field extends into the *n*-type bar from the *p*-type regions. This charge focuses current carriers (electrons) away from the region, reducing the amount of bar available for conducting the current between the source and drain (shown diagrammatically in Figure 11.6).

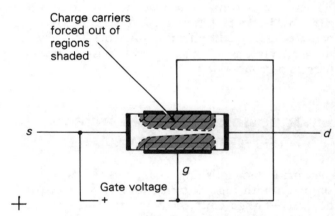

Figure 11.6 The electric field forces charge carriers away from the plates when a negative potential is applied to the gate.

Figure 11.4 Theoretical construction of an *n*-channel depletion-mode FET.

A bar of *n*-type semiconductor (almost invariably silicon) is made with shallow *p*-type regions in the upper and lower surfaces. These are connected to the gate terminal, and the two ends are connected to the source and drain. If the bar is connected to a voltage source, current will flow through it. Since the bar is symmetrical, it can flow either way, the source and drain being interchangeable. The current flow consists of electrons moving through the *n*-type semiconductor (Figure 11.5).

If the potential applied to the gate is made sufficiently negative, the electric field will extend across the whole thickness of the bar of *n*-type semiconductor and hardly any charge carriers will be available for current flow. The current available from the drain will drop to a very low value (but never quite to zero; the channel never closes completely – see Figure 11.7).

Changes in the voltage applied to the gate will cause corresponding changes in the current flowing between the source and the drain, which makes the

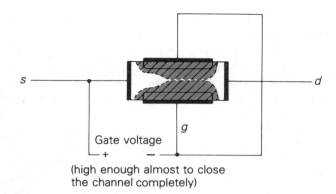

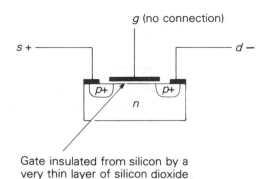

Gate voltage
+   −

(high enough almost to close
the channel completely)

Figure 11.7 Although the current flow can be greatly reduced, it is impossible to stop it completely. There will always be a small gap between the two areas that are without charge carriers.

g (no connection)

s +   d −

Gate insulated from silicon by a
very thin layer of silicon dioxide

Figure 11.8 Theoretical construction of an *n*-channel enhancement mode FET. The gate is completely insulated from the rest of the structure by a very thin layer of silicon dioxide.

operation of the FET very similar to that of a bipolar transistor *except* that there is no gate (base) current: remember that the FET is a voltage-controlled device.

Insulated-gate field-effect transistors

More generally known as a MOSFET (**metal-oxide semiconductor field-effect transistor**), the insulated-gate FET is one of the most important devices in the electronics industry. There are two basic categories of MOSFET, known as **depletion MOSFETs** and **enhancement MOSFETs**. They both work on a different principle from the JUGFETs that we have been looking at so far in this chapter.

Enhancement MOSFETs

The structure of a *p*-channel enhancement **MOSFET** – once again, only a theoretical structure – is shown in Figure 11.8. The most striking feature is the gate. It is insulated from the silicon by a thin layer of silicon dioxide. The layer is very thin, typically only about 0.1 μm thick. Although very thin, this silicon dioxide layer has an enormously high resistance: at least 10 GΩ.

The *n*-type silicon has two regions of heavily doped *p*-type impurity, connected to the source and drain leads. With no applied gate voltage, one of the *p–n* junctions will act like a reverse-biased diode and block any flow of current. If a negative

potential is applied to the gate electrode, holes from the *p*-type regions are attracted into the area immediately beneath the electrode. This effectively, if temporarily, makes a narrow *p*-type region just beneath the gate (Figure 11.9). The blocking *p–n* junction is bypassed by this induced channel of *p*-type material, and electrons can flow through the device, between the source and drain.

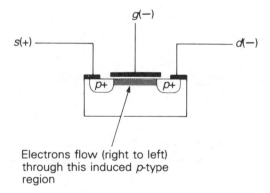

g(−)

s(+)   d(−)

Electrons flow (right to left)
through this induced *p*-type
region

Figure 11.9 A *p*-type region is induced in the *n*-type bar when the gate is made negative.

The circuit symbol for the *p*-channel enhancement MOSFET is given in Figure 11.10. The centre connection in the circuit symbol (with the arrowhead) is a connection to the silicon substrate (the chip itself). Often the connection is made internally to the source, but sometime manufacturers fit a fourth lead so that the **substrate** (the slice of silicon that the device is built on) can be used as a second 'gate'. The conduction characteristics of the device then depend approximately on

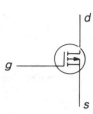

Figure 11.10 The circuit symbol for a *p*-channel enhancement MOSFET.

the difference in potential between the gate and substrate.

When the circuit symbol has an arrowhead pointing away from the gate, it symbolises a *p*-channel device. It is, however, equally possible to make an *n*-channel device, with all the polarities reversed, using *n*-regions diffused into a *p*-type bar. In this case the charge carriers are electrons attracted from the *n*-regions; otherwise operation is the same. The circuit symbol for an *n*-channel enhancement MOSFET is given in Figure 11.11.

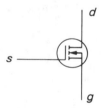

Figure 11.11 The circuit symbol for an *n*-channel enhancement MOSFET.

Depletion MOSFETs

The second class of MOSFET is the **depletion MOSFET**. The structure is shown in Figure 11.12; it is very similar to that of the enhancement

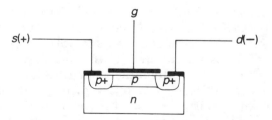

Figure 11.12 Theoretical structure of a depletion MOSFET.

MOSFET. Notice that there is an 'extra' region. A narrow strip of *p*-type impurity has been diffused into the space below the gate, so that the depletion MOSFET, with no signal applied to the gate, looks rather like the enhancement MOSFET when its gate is connected to make it conduct. Compare Figures 11.12 and 11.9.

Applying a positive voltage to the gate causes electrons from the *n*-type region to be attracted to the area under the gate electrode, neutralising some of the holes in the *p*-type channel and reducing the amount of current flowing between the source and drain. The higher the positive potential (for a *p*-channel device, of course), the more the source–drain current is cut off.

The depletion MOSFET can also be used in the enhancement mode as well.

Applying a negative voltage to the gate of a *p*-channel device will increase the source–drain current by adding to the number of holes available as charge carriers.

Figure 11.13 shows the circuit symbols for *p*-channel and *n*-channel depletion MOSFETs.

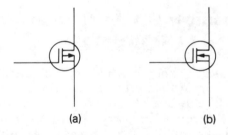

(a)   (b)

Figure 11.13 The circuit symbols for (a) *p*-channel and (b) *n*-channel depletion MOSFETs.

The importance of MOSFETs

It seems paradoxical that, even though the MOSFET is a vitally important device, individual MOSFETs are quite rarely used. Only in the highest-quality communications receivers do we find MOSFETs used as a matter of common practice. MOSFET technology is used overwhelmingly in **integrated circuits**. We shall take a brief look at these in the next chapter.

Before leaving FETs (for the time being), let us

just list some of the more interesting aspects of the devices.

1. Because of the necessity for a very narrow channel, it is difficult to make FETs that will carry high currents. Most (but not all) FETs are therefore low-power devices.
2. Very high-frequency operation of FETs is hampered by the internal capacitance, a by-product of the very narrow regions. There is an effective capacitance of a few picofarads between the source (and drain) and the gate.
3. It is possible to make FETs *very* small.
4. Because of the very high input resistance and the thinness of the silicon dioxide insulating layer, MOSFETs are very liable to damage from high voltages accidentally applied between the gate and other terminals. For this reason, most MOSFETs are made with a Zener diode connected between the gate and substrate. Normally, the Zener diode is non-conducting, but if the gate voltage rises too high, it conducts and discharges the gate. Even so, MOSFETs are easily damaged by static electricity. The charge on a person wearing shoes with rubber or plastic soles can rise, on a dry day, to several kilovolts, enough to destroy the MOSFET, Zener and all. Just touching the gate terminal can destroy the device!

The terminals of a MOSFET are generally shorted together with metal foil or conductive foam until it is installed in a circuit. Extra care is needed when servicing any equipment that might include MOSFETs.

■ CHECK YOUR UNDERSTANDING

● **Unipolar transistors** (field-effect transistors, or FETs) are voltage-operated (bipolar transistors are current-operated).
● In a unipolar transistor a varying voltage on the **gate** controls the amount of current that can flow between the **source** and **drain**.
● There are two major types of FET: **junction gate** (JUGFET) and **insulated gate** (IGFET, more often known as a MOSFET).
● JUGFETs have a high input resistance, but IGFETs have a very high input resistance, measured in gigohms.
● *n*-channel and *p*-channel FETs are available.
● Enhancement-mode and depletion-mode FETs are manufactured.
● FETs are damaged by high voltages, even electrostatic voltages. It is essential to take precautions when handling these devices, to ensure that they are never exposed to static electricity.

REVISION EXERCISES AND QUESTIONS

1 Draw the circuit symbol for an *n*-channel FET.
2 What is the principal difference between a bipolar transistor and a FET?
3 What special precautions must be taken when handling FETs?
4 MOSFETs are very important devices as they are used in . . . ? Complete the sentence.

Integrated circuits and semiconductor devices

Introduction

In this chapter I shall be considering – all too briefly at this stage in your studies – the fascinating world of integrated circuits, the aspect of electronics that has the greatest impact of all.

Integrated circuits

Semiconductor manufacture, or 'fabrication' as it is more generally called in the industry, is a highly specialised and very difficult subject, and may not be included as part of your syllabus. However, it is useful for a student of electronics to have at least some idea of the principles involved, particularly as it goes some way to explaining what integrated circuits are all about.

Various different semiconductor materials are used as a basis for semiconductor manufacture, but the most common (and cheapest) is **silicon**.

The first step is to take a single large crystal of pure silicon. This is not as easy as it seems, for silicon that only 30 years ago would have been called 'chemically pure' would be hopelessly contaminated for the purpose of semiconductor manufacture. The silicon crystal is sliced up (like a sausage) into circular **wafers**, typically 50 mm diameter and about 0.5 mm thick. The surfaces are ground and polished perfectly flat, leaving the wafer about 0.2 mm thick. The wafer is finally cleaned with chemical cleaners.

The technique for making a single *pn* diode is as follows. Beginning with a wafer of *p*-type silicon, made by adding a tiny amount of *p*-type impurity such as indium or boron to the pure silicon, an *n*-type **epitaxial layer** about 15 μm thick is 'grown'

on the wafer. This is done by heating the wafer to about 1200 °C in an atmosphere of silicon and hydrogen tetrachloride with a trace of antimony, phosphorus or arsenic. Next a thin (about 0.5 μm) layer of **silicon dioxide** is grown on top of the wafer by heating it to about 1000 °C in an atmosphere of oxygen or steam. Silicon dioxide has three very useful properties. First, it is chemically rather inert and is not attacked by gases in the atmosphere, or indeed most other chemicals, even at high temperatures. Second, it is impervious and prevents the diffusion of impurities through it. Third, it is an excellent electrical insulator. The wafer at this stage is shown in cross-section in Figure 12.1.

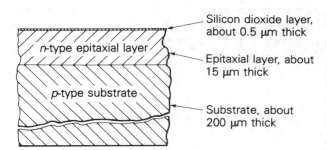

Figure 12.1 A silicon wafer in cross-section, after formation of a silicon dioxide layer on its surface.

Next, openings have to be made in the silicon dioxide layer, in the right places. This is done photographically. The wafer is coated with a **photoresist**, a light-sensitive emulsion similar to the emulsion on a black-and-white film. A pattern, or mask, is removed, and the wafer is washed with trichloroethylene, a chemical that dissolves the photoresist only in places where it was not exposed to ultraviolet light. This leaves the photoresist as a 'negative' of the mask.

At the end of this stage, the wafer is washed in

hydrofluoric acid: a very powerful acid that will even dissolve glass, but not the special photoresist (nor, fortunately, the silicon wafer!). The wafer is washed again and the resist is removed with hot sulphuric acid. The various steps are shown in Figure 12.2.

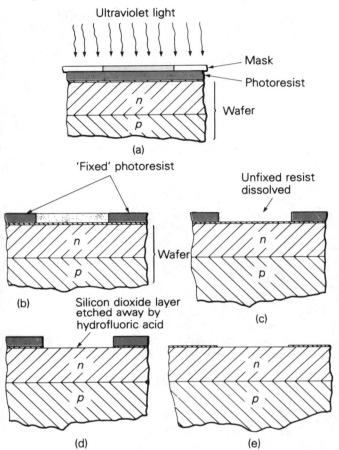

Figure 12.2 The various stages in masking and etching the silicon dioxide layer.

The end result of this is to produce the right pattern of holes in the silicon dioxide layer. The wafer now goes back in the furnace, with an atmosphere of p-type impurity. The impurity diffuses into the wafer below the holes in the silicon dioxide, but nowhere else (Figure 12.3).

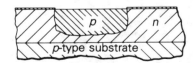

Figure 12.3 p-type impurity is diffused into the upper n-type layer.

The whole process is repeated, with a different mask, to diffuse another n-region into the p-region just created. Then the whole process is repeated a third and then a fourth time, to diffuse more doped regions into the wafer. Figure 12.4 shows the final result. Connections have been made to the heavily doped $n+$ regions, by evaporating an aluminium film onto the wafer, after suitable masking. And that's the diode finished.

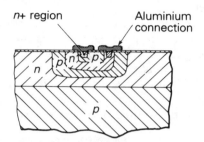

Figure 12.4 A completed silicon planar epitaxial pn diode.

Notice that those extra p and n layers below the diode proper (the p–n junction that is connected to the outside world) serve to isolate the diode from the rest of the wafer: a reverse-biased p–n junction surrounds all parts of the diode, preventing leakage to the rest of the wafer. This isn't at all useful if we are going to cut the wafer up into individual diode chips. But suppose we cut the wafer into larger sections with several diodes on each? The aluminium layer could carry current from one part of the wafer to another, and transistors and diodes could all be combined to build a complete circuit!

The first integrated circuits used just this technique, though more recent devices are rather more subtle, and rely on electrical connections inside the silicon structure rather than on superimposed aluminium 'wiring'.

Diodes and transistors – bipolar and FETs – can be produced on a silicon wafer. So, too, can resistors: they are made from **tantalum** (a poor conductor) deposited on top of the wafer, or are built into the wafer as a 'pinch' resistor. The pinch effect is similar to that observed in a partially turned-off FET and relies on a very thin region for conduction. Resistance values of up to 100 kΩ can be produced in this way.

Although it is possible to make capacitors on an integrated circuit, the values are generally limited to a few picofarads if the capacitor region is not to be excessively large. But big capacitors can generally be avoided in integrated circuits.

There is no equivalent of an inductor, but fortunately most circuits can be designed without inductors.

Once the wafer has been tested, it is cut up and any faulty circuits discarded, again automatically. The chip is mounted on a suitable frame, and connections are made between the aluminium 'pads' on the chip and the frame that forms the connections to the chip. Fine gold wire is often used for the connections.

It remains only to put the integrated circuit in a suitable container: 'encapsulation' is the word most often used. Various styles of encapsulation are used, but by far the most common is the DIL pack (dual-in-line). DIL packs are based on a standard 0.1 inch matrix. The pins down each side are always 0.1 in apart, and the distance between the two rows is a multiple of 0.1 in, generally 0.3 in for packages of 18 pins or less, and 0.6 in for 20 pins or more. Figure 12.5 illustrates typical plastic DIL packs.

Figure 12.5 A selection of integrated circuits in DIL packs, one of the most common forms in which integrated circuits are made.

Figure 12.6 An example of a large-scale integrated circuit. This IC contains a complete computer central processor.

The number of individual semiconductor devices that can be packed onto a single chip is astonishing. A 40-pin DIL pack can contain the electronics for a complete computer! Figure 12.6 shows a moderately complex integrated circuit (the central processor unit of a simple computer). Devices like this have a huge number of transistors and diodes, approaching 10 000 on a chip less than 10 mm square.

> The integrated circuit is the key to modern microelectronics. It's what makes a study of semiconductors and semiconductor physics exciting and relevant. And it's why modern electronic systems – from personal stereo players to defence radar installations – can do so much, so well, so cheaply.

Semiconductor devices

There are a number of semiconductor devices – other than bipolar and unipolar transistors – that are in common use. This section lists them and gives short descriptions of what they do and what they are for.

Thyristors

The thyristor, also known as a **silicon-controlled rectifier** (SCR), is a component that has wide applications in the field of power control. In essence it is very simple, easy to use and easy to understand. The circuit symbol for the SCR is given in Figure 12.7.

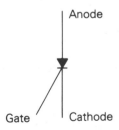

Figure 12.7 The circuit symbol for a thyristor (SCR).

As you can tell by looking at the symbol, the SCR is a form of diode. Under normal circumstances, however, it will not conduct current in either direction. If a voltage is applied to the SCR in such a way that, if it were an ordinary diode, it would conduct (forward-biasing it), a small current made to flow between the **gate** and the **cathode** will make the SCR change abruptly from a non-conducting to a conducting state. It then has characteristics the same as those of a forward-biased silicon diode, with a somewhat higher forward voltage drop (typically 0.7–1.3 V).

Turn-on takes place extremely rapidly, within a few microseconds of the application of the gate current. Once it has been turned on, the SCR will stay in the conducting state even if the gate current is stopped. It will turn off again only when the current flowing through it is reduced below a certain (quite low) value. This minimum current required to maintain conduction is called the **holding current**, and is between a few microamps and a few tens of milliamps, according to the type of device.

SCRs can be made to withstand high voltages, and are readily obtainable for use with peak voltages from 50 V to tens of kV.

SCRs are **power-control** devices, and are made in a variety of current-carrying capacities, from small devices capable of handling a few hundred milliamps at 50 V to huge units designed to deal with hundreds of amps at hundreds of volts. Small SCRs used in electronic circuits likely to be encountered by the engineer range from T092 encapsulated 300 mA versions up to about 40 A. Voltages are usually in the range 50–1200 V, with 400 V types preferred for control of mains electricity at 220–240 V.

The size of the **gate current** – the minimum amount needed to trigger the SCR – can range from a few hundred microamps in sensitive low-voltage SCRs to a few tens of milliamps for the larger ones.

The fact that a thyristor, once triggered, remains on until the current flowing through it is interrupted does not necessarily mean it can't be used in d.c. circuits. The circuit in Figure 12.8 illustrates how it can be turned on and off to control a relay, motor or lamp. Momentarily connecting T_1 to the positive line switches the thyristor on by providing a gate current. When the thyristor is on, the 1 μF capacitor charges up through the 1 kΩ resistor, so that its left side (in the diagram) is positive.

The thyristor can be turned off by momentarily connecting T_0 to the positive line. Until the capaci-

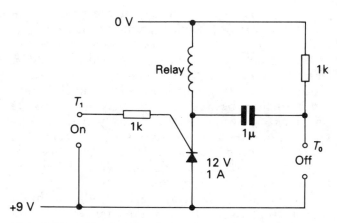

Figure 12.8 *A thyristor circuit to give 'on' and 'off' control.

tor has discharged (which takes less than a millisecond) this provides a voltage that opposes the direction of current flowing through the thyristor, and the current drops to zero for long enough to turn it off.

This circuit will drive any suitable relay or lamp.

Triacs

The triac, to give the more usual (but less informative) name for the **bidirectional thyristor**, is an SCR that will conduct **in either direction**, with the gate current flowing between **anode 1** and the **gate** in **either** direction: a very accommodating component! The circuit symbol is given in Figure 12.9.

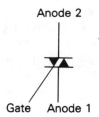

Figure 12.9 The circuit symbol for a triac.

Like the SCR, the triac will remain on, once triggered, until the current passing through it falls below the level of the holding current. Triacs are often used with mains electricity, for controlling the brightness of electric lamps. When used with the mains a.c. supply, the triac will automatically turn off 100 times a second with a 50 Hz supply frequency. You should be able to explain why!

Diacs

The diac is a specialised component, specifically designed for use with SCRs and triacs, though it has inevitably found its way into various other circuits. The 'long' name for the diac is the **bidirectional breakdown diode**, and its circuit symbol is given in Figure 12.10.

Figure 12.10 The circuit symbol for a diac.

A diac normally blocks the flow of current in either direction, but if the voltage across it is increased to the **breakover voltage**, usually about 30 V, the diac begins to conduct. It does in fact exhibit a phenomenon known as **negative resistance**, for as breakover occurs the voltage across the diac actually drops by a few volts. If the diac is connected in a circuit in which a steadily increasing voltage appears across it, it will, at breakover, allow a sudden current 'step' to flow in the circuit. Figure 12.11 shows this in graphical form.

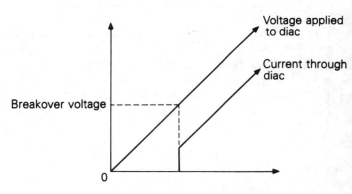

Figure 12.11 Graph of voltage and current for a diac.

The diac is an ideal device for providing a suitable trigger pulse for an SCR or triac. The circuit in Figure 12.12 is a typical triac/diac application: a circuit for a mains-lamp brightness controller.

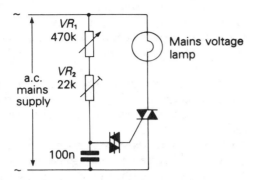

Figure 12.12 A simplified but functional circuit to control the brightness of an a.c. lamp.

> ⚠ This simple circuit is not recommended as a construction project without supervision because mains voltages are present and stringent safety precautions must be taken.

The circuit works by means of what is loosely known as 'phase control'. As the voltage across the circuit rises, following the sine wave of the mains electricity supply, the capacitor charges up at a rate controlled by VR_1 and VR_2. At a certain point in the a.c. half-cycle the voltage across the capacitor will reach the breakover voltage of the diac; the diac applies a pulse of current to the triac gate, triggering the triac into conduction and allowing current to flow through the load. At the end of the mains supply half-cycle the triac will switch off.

During the next half-cycle, the same sequence will occur, but with all the polarities reversed (unimportant, because all the circuit components will work with a voltage applied in either direction). Once again, the triac will trigger after a certain delay, allowing a current flow through the load.

If the resistor VR_1 is set to a high resistance, the capacitor will charge up slowly and the triac will be triggered near the end of each half-cycle. If the resistor VR_1 is set to a low resistance, the triac will be triggered near the beginning of each half-cycle, applying almost full power to the load. The amount of power flowing through the load is controlled by VR_1, so the brightness of a lamp can be regulated over a wide range. The graph in Figure 12.13 illustrates this. Because the circuit works by switching the power on and off, little heat is dissipated and the control is efficient in terms of power used. VR_2 is included in this circuit to set the maximum

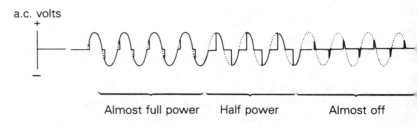

Figure 12.13 A graph illustrating the way in which the triac can be used with an a.c. supply to control the amount of power flowing through a load.

brightness level according to the characteristics of the diac and triac.

Unijunction transistors

Unijunction transistors (UJTs) are used primarily in oscillators: circuits that produce a steady-state alternating voltage or current at a particular frequency and amplitude. We shall look at oscillators in detail in Chapter 15. A typical unijunction oscillator is shown in Figure 12.14.

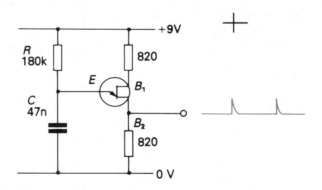

Figure 12.14 *A unijunction oscillator.

The UJT has three terminals: two **bases** and an **emitter**. The UJT in the circuit is an *n*-channel device; *p*-channel UJTs are available but uncommon (the circuit symbol has the arrowhead reversed for the *p*-channel version).

The UJT normally exhibits a high resistance between the two bases: about 10 kΩ. If the emitter is at a low potential relative to B_1, the emitter–B_1 junction behaves like a reverse-biased diode, and a negligible current flows through the emitter circuit. If the emitter potential is increased to a point approximately equal to half the voltage between B_1 and B_2, the emitter–B_1 junction suddenly becomes forward-biased and the emitter–B_1 resistance falls to a very low value.

In the oscillator circuit illustrated, the timing is controlled by R and C. When the UJT conducts, the capacitor is discharged very rapidly through the emitter–B_1 junction and the 820 Ω resistor. An output taken from B_1 therefore consists of a series of short pulses at the required frequency.

As we shall see later, there are many kinds of oscillator. The unijunction oscillator is just one of them.

■ CHECK YOUR UNDERSTANDING

● **Integrated circuits** are usually made on **chips** cut from thin **wafers** of very pure silicon.
● An *n*-type region is made by heating a masked area of the silicon wafer to around 1200 °C in an atmosphere containing trace elements that result in *n*-type impurities.
● A pattern of silicon dioxide is produced on the surface, and the chip is reheated in an atmosphere containing elements that will make *p*-type impurities.
● Various *p*-type and *n*-type regions and layers can be built up to form a variety of different semi-conductor devices, linked together in a pattern that makes a complete circuit.
● A **thyristor** (silicon-controlled rectifier) is like a diode, but is non-conducting until turned on by a current flowing between its **gate** and **cathode**. Once turned on, it will remain on until the current flowing through it (from anode to cathode) drops to near zero.
● A **triac** works like two thyristors – in parallel and facing in opposite directions – that share a common gate. A triac is used to control **alternating** currents.
● A **unijunction transistor** is a device that has a high resistance between its two **base** connections until the voltage on its **gate** reaches about half the potential between the bases. The resistance between the bases then drops to a low value. Unijunction transistors are commonly used in oscillators.

REVISION EXERCISES AND QUESTIONS

1 What is the material normally used for the wafer in integrated circuits? Why?
2 Draw a circuit in which a thyristor is used to switch on a LED, and a mechanical switch is used to switch off power to the whole circuit.
3 What is a triac? Describe its operation.
4 Class discussion: 'Integrated circuits have had more impact on civilisation than any other technological advance this century. True or not?'

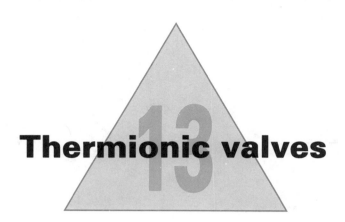

Thermionic valves

Introduction

This short chapter is about valves. Americans call them 'tubes'. Valves are not used in 'consumer' electronics any more, but are still found in high-power radio and television transmitters: not portable equipment, but the kind that broadcasting stations use. Valves are useful when a lot of power at high voltage is needed. It is interesting to look at valves in terms of the history of their development.

Thermionics

Thomas Alva Edison is chiefly known as the inventor of the phonograph and as one of the inventors of the electric lamp. It is less well known that Edison *nearly* invented the thermionic diode: the 'breakthrough' device that heralded the beginning of the science of electronics. The word 'thermionics' is from 'thermion', an obsolete word used to describe a free electron that has been released by heating.

In the early 1880s Edison was trying to improve his electric lamp. One of the problems he had was in preventing the inside of the glass envelope going black and obscuring the light. He realised that the filament of the lamp was evaporating, and wondered if he could use a wire grid to intercept the material before it got to the glass. The modified lamp was based on the design in Figure 13.1.

Unfortunately it didn't work, but during his experiments Edison did try the effect of applying an electric voltage to the grid. He observed that if the grid were made negative with respect to the filament, nothing much happened, but if he made

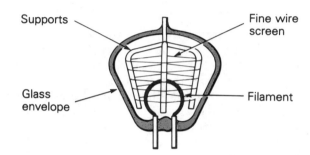

Figure 13.1 Edison's modified lamp.

it positive with respect to the filament, he could draw a substantial current from it. This was mildly interesting but was not going to help him with the electric lamp, so he called it the 'Edison effect' and filed the idea away.

It was left to John Ambrose Fleming, in 1904, to do the essential pioneering work in electronics, and the device that carries his name, the Fleming diode, bears tribute to his work.

Thermionic diode

The simplest form of Fleming diode (di- from the Greek for 'two') is almost identical to Edison's modified lamp, and is illustrated in Figure 13.2. The cathode is just like a lamp filament, though it is treated with thorium to increase the number of electrons that can be driven off when it is heated. Facing the cathode is the anode, a metal plate connected to an outside terminal. The two parts are sealed inside a glass envelope in which there is a vacuum. It is easy to see how the Fleming diode, or thermionic diode, works.

When the cathode filament is heated by passing

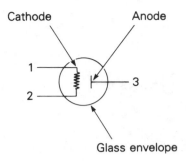

Figure 13.2 A Fleming diode.

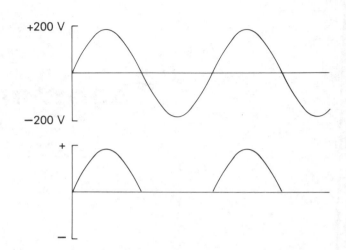

Figure 13.4 Rectification of an alternating voltage.

a current through terminals 1 and 2 in Figure 13.2, electrons are driven off the filament. These free electrons (thermions) are separated from their parent atoms by the heat. If the anode is made negative compared with the cathode, the electrons are repelled by the anode (like repels like; remember that the electrons have negative charges) and no current can flow through the diode.

However, a positive charge on the anode will attract electrons given by the cathode, and they will move through the vacuum inside the envelope; this movement of electrons constitutes an electric current. Figure 13.3 shows a demonstration circuit for the thermionic diode.

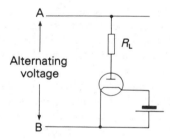

Figure 13.3 A demonstration circuit for a thermionic diode.

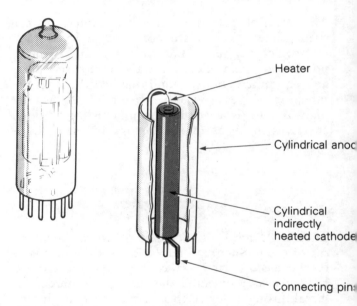

Figure 13.5 Construction of a thermionic diode valve.

The single cell provides heating current for the cathode, but otherwise plays no part in the operation of the circuit. Consider the effect of an alternating voltage applied across A–B in Figure 13.3. While A is positive with respect to B a current will flow through the load resistor R_L, but if B is more positive than A, no current will flow. Figure 13.4 compares the applied voltage with the current flowing through R_L.

This is the principle of rectification: turning a.c. into d.c. The original Fleming diode was developed into the **rectifier valve**, shown in Figure 13.5. Note that the cathode is **indirectly heated**. This design separates the heating and electron-

emitting functions. The heater filament is inside a narrow tube of thin metal, but nowhere does it come into contact with the tube. The filament gets red hot, and this heats the tube (the cathode) to around 1750 °C, at which temperature it emits electrons. The reason for this is to isolate the heater from the cathode, both electrically and thermally. It means that the valve heaters can all be run from the same power supply, which can be a.c.; although the temperature of the heaters will fluctuate (at 50 Hz) with the changing current of the supply, that of the cathodes will remain relatively constant, because they take a little while to heat up or cool down. Running the heaters from a.c. meant that mains-operated radio receivers were possible. Before that, 2 V lead–acid accumulators were used to power valve heaters, and 120 V dry batteries – 80 Leclanché cells in series! – for the anode and grid voltages.

Modern valves for powerful transmitters use directly heated cathodes, powered by stable d.c. power supplies.

Another obvious (but important) development was the idea of putting all the connections at one end of the valve, so that it could be plugged into a base for easy replacement when the heater burned out.

Thermionic diodes like the one in Figure 13.5 were in regular use in all types of equipment until the end of the 1960s, when they began to be superseded by solid-state diodes. Valves are no longer used at all, except in some very specialised applications such as very high-powered radio transmitters.

Thermionic triode

The triode (tri- from the Greek for 'three') valve was invented by Lee De Forest in 1910. It is constructed like a Fleming diode but with an extra electrode, consisting of a **grid** of fire wires, interposed between the cathode and the anode. The basic layout is shown in Figure 13.6, which, incidentally, is the circuit symbol for the triode valve.

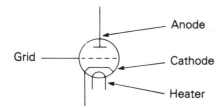

Figure 13.6 The circuit symbol for a thermionic triode.

If the control grid is left unconnected the triode behaves like a diode, but if the grid is kept at a small negative potential relative to the cathode, the current flowing through the valve is reduced. It is easy to see why this is so; electrons travelling from the cathode to the anode are repelled by the charge on the grid and prevented from passing. If the grid is made slightly more negative still, the current flow from the anode ceases altogether.

For all practical purposes the grid takes no current at all, and the maximum voltage required on the grid to stop all current flow from the anode is substantially less than the anode voltage. Between certain limits (which depend on the valve construction and on the circuit in which it is used) the voltage at the anode of the valve is proportional to the grid voltage, but larger. Thus the valve can be used to amplify a signal applied to the grid, by increasing the amplitude (size) of the signal without changing its form.

In operation the thermionic triode is therefore similar to a FET, but with the following disadvantages: it is thousands of times bigger; it is thousands of time less reliable; it wastes tens of thousands of times more power; it is much more expensive to make. But, of course, valves were in use years before solid-state electronics had even been considered.

Figure 13.7 shows a triode valve in a simple amplifier circuit. The two resistors R_g and R_c are involved in keeping the grid at a more negative potential than the cathode. The output waveform as shown in the figure is the same shape as the input waveform, but has a larger voltage change: it is amplified. A small voltage applied to the grid controls a much larger voltage.

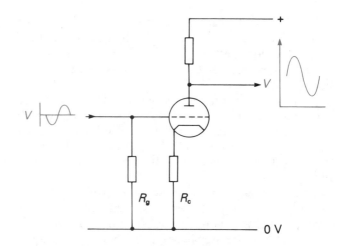

Figure 13.7 A triode valve amplifier.

There were many types of valve apart from the triode. Almost all the designs are intended to improve the basic efficiency or to compensate for some undesirable characteristic. Almost all these improvements made the valve more complicated and therefore more expensive.

Thermionic tetrode

The tetrode valve (tetra- from the Greek for 'four') is an improvement on the basic triode design, and has an additional grid – called the **screen** – which is positioned between the control grid and the anode (Figure 13.8). The screen is kept at a steady positive voltage. This has the effect of increasing the gain of the valve, giving far greater amplification than can be obtained with a triode. In the triode valve, the anode current is dependent on the anode voltage. As the anode voltage swings up and down to follow a varying input signal, the changing voltage on the anode has the effect of reducing the gain.

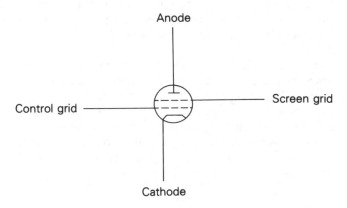

Figure 13.8 A tetrode valve.

In the tetrode valve, the presence of the screen means that the anode current is less dependent on the anode voltage than it is in a triode. The screen voltage is set so that it is just less than the 'no signal' anode voltage. The screen also has the important function of reducing the capacitance between the control grid and the anode, which enables the valve to be used for higher-frequency signals. Various aspects of the mechanical design of the tetrode are aimed at reducing this capacitance as far as possible.

The tetrode is in many ways more useful than the equivalent triode, but the screen grid causes its own problem: when the anode voltage swing is such that on the negative-going peak of a signal the anode voltage falls below that of the screen, electrons flow from the anode to the screen. This prevents the valve from being used in applications where a large output signal (and high output power) is needed, as at this point the onset of distortion is very rapid.

To solve this problem, yet another grid has to be added to the valve.

Thermionic pentode

The pentode valve (penta-, as I expect you guessed, from the Greek for 'five') is like a tetrode, but with an additional grid between the screen and the anode (Figure 13.9). This grid is usually connected to the cathode. Its purpose is to interpose a low voltage between the anode and the screen, to prevent the anode–screen current flow. The pentode can be operated with much bigger anode voltage swings without distortion.

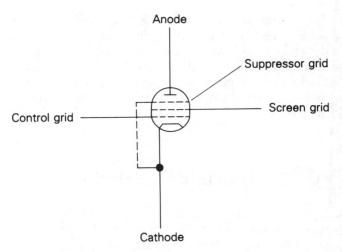

Figure 13.9 A pentode valve.

Figure 13.10 shows an audio amplifier based on a pentode valve. The circuit looks quite simple, but bear in mind that the output transformer that connects it to the speaker is likely to be quite large; the circuit requires a 200 V d.c. power supply, and also a low-voltage heater supply for the valve (which rules out battery operation). It gives less amplification, more distortion, is less reliable, and

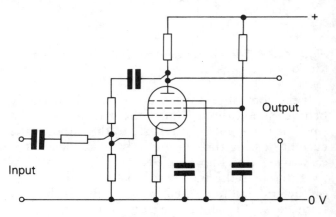

Figure 13.10 A pentode valve amplifier circuit.

wastes vastly more power than the LM380 solid-state amplifier in Chapter 14.

CHECK YOUR UNDERSTANDING

● **Valves** are 'vacuum-state' devices – now obsolete except in powerful radio and television transmitters – that can be used to amplify a signal in much the same way as a field-effect transistor.
● Valves require a **heater** to heat up the cathode and drive electrons off it. The heater wastes a lot of power.
● Valves operate on relatively high voltages compared with solid-state devices.
● Valves are much bigger and less reliable than solid-state devices like transistors.
● **Diode** valves rectify a.c.

● **Triode** valves are basic amplifiers, with three connections, much like FETs.
● **Tetrode** valves are like triodes but with an extra grid: the screen grid.
● **Pentode** valves are like tetrodes but with a second screen added.

REVISION EXERCISES AND QUESTIONS

1 Sketch the circuit symbol for a pentode valve.
2 What is the purpose of the **heater** in a valve?
3 Where will you find electronic valves in use today?
4 Describe, in one sentence, the basic principle of the triode valve.

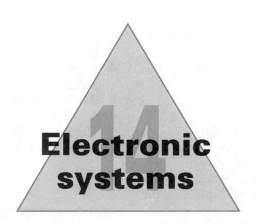

Electronic systems

Introduction

The next two chapters of this book will begin to deal more with practical matters: with systems and practical circuits rather than with theory and electronic devices. They also introduce much more project work, for which the preceding chapters have been an essential preparation.

Power-supply circuits

The provision of a stable direct voltage is an important part of circuit design. In electronics, a 'power supply' isn't usually a primary source of energy; it is a circuit that converts mains power into a source of voltages, with various current capabilities, that drive the rest of the circuits in a piece of equipment.

A.C. to d.c. conversion

The first step is usually the conversion of a.c. mains to a lower voltage. This is done with a **transformer**. A purpose-built mains transformer should be used, as this will have an insulation resistance between the two windings that is sufficiently high to satisfy national electrical safety regulations. It is essential that mains power cannot be conducted to the secondary winding, as this would result in the equipment becoming 'live'.

Diodes are used to rectify the alternating current from the transformer secondary to make it into d.c. Figure 14.1 shows a transformer driving a lamp with a.c. Compare the waveform across the lamp with the one in Figure 14.2. The diode has blocked

current flow in one direction and the bulb is receiving only alternate half-cycles of the supply. Four diodes can be used in the **bridge rectifier** configuration to make use of positive and negative excursions of the a.c. waveform (Figure 14.3).

To smooth the output of the bridge rectifier, the simplest power-supply circuit uses a large-value capacitor connected across the output. A capacitor

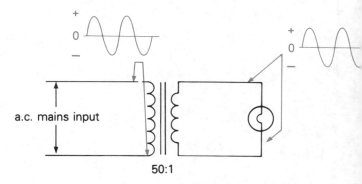

Figure 14.1 The waveforms at the input and output of a transformer driving a lamp.

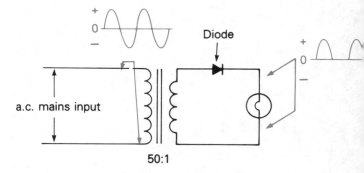

Figure 14.2 The waveforms for half-wave rectified direct current, supplied by the secondary of the transformer, through a diode.

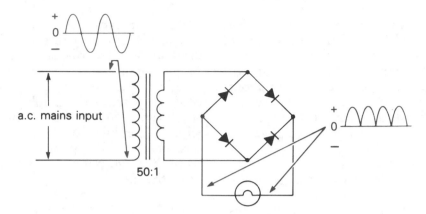

Figure 14.3 The diode bridge rectifier, used to provide full-wave rectification of an a.c. supply. Although both halves of the a.c. input waveform pass through the lamp, the current flow through the lamp is not smooth, and this would upset the operation of many circuits.

used in this kind of circuit is called a **smoothing** or **reservoir capacitor**. The effect of the capacitor is shown in Figure 14.4. The capacitor supplies the current required by the load when the voltage drops between each half-cycle of the mains supply. When the voltage rises, the capacitor is recharged, so that it has energy available to fill in the 'gaps' in continuity of the power supply to the circuits.

For many circuits, this kind of power supply is sufficient. If better voltage regulation (evenness of voltage) is essential, then a **regulator circuit** is connected between the basic power supply and the circuits being supplied.

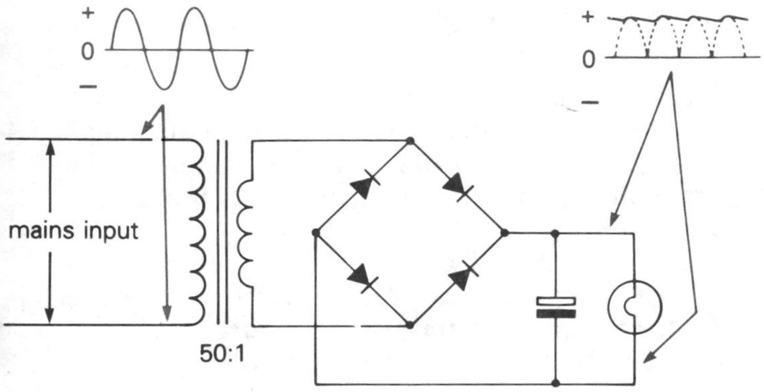

Power-supply regulators

In Chapter 9, I mentioned the Zener diode as a component that is useful for providing a reference voltage. Many circuits require a very stable supply voltage, and the Zener diode is generally used in this context. A useful circuit is given in Figure 14.5.

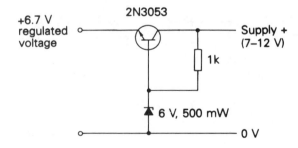

Figure 14.5 *A simple voltage regulator using a Zener diode.

This circuit uses a bipolar transistor to carry the load current, and also to amplify the regulating effect of the Zener diode to provide better regulation. Operation of the circuit is as follows. If there is no load, no current flows through the collector–emitter junction of the transistor. The only current is that flowing through the resistor and the diode: just a few milliamps.

When a load is connected, current flows through the transistor, since the base is held at (in this case) 6 V, which biases the transistor into conduction. All the while the voltage at the emitter is at or

Figure 14.4 A bridge rectifier with a smoothing capacitor. A large capacitor is used to provide power during the gaps in the rectified waveform.

below 6.7 V (allowing for the voltage drop across the junction) the transistor remains forward-biased. But the voltage at the emitter can never rise above 6 V, as this would involve a higher voltage on the emitter than on the base, reverse-biasing the junction and causing the transistor to switch off. In practice, the voltage on the emitter rises to about 6.7 V and stays there. If the input voltage to the regulator changes, this does not affect the regulated voltage unless, of course, the input voltage drops below 6.7 V.

Similarly, changes in the resistance of the load will result only in a compensating change in the bias conditions of the transistor; the voltage at the emitter remains substantially constant. This type of regulator is useful and inexpensive. The output current is limited by the current and power rating of the transistor.

It is convenient to fit the whole of the regulator circuitry on an integrated circuit, and such components are available cheaply. Figure 14.6 shows a 5 V regulator circuit, capable of handling currents up to 1 A. The performance of the IC is considerably superior to that of the circuit in Figure 14.5, with more accurate regulation, short-circuit protection and automatic shut-down in the event of the IC overheating.

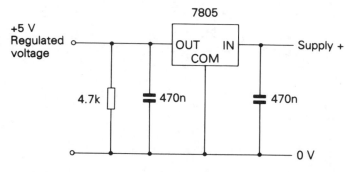

Figure 14.6 *A high-performance 5 V regulator using an integrated circuit.

Voltage regulator designs are many and varied, but these days most are based on ICs.

Transistor amplifiers

You saw in Chapters 10 and 11 how transistors – bipolar and FET – work as amplifying devices, using a small current or voltage to control a much larger current. The application to a machine such

as a record-player, for example, is an obvious one, for the small electric signal produced by the player's pick-up cartridge must be amplified to a sufficiently large extent to drive a speaker. A system diagram of a record-player amplifier looks quite simple (Figure 14.7). The symbols for the cartridge, amplifier and speaker are those conventionally used. In practice the amplifier breaks down into two main sections: the preamplifier, which deals with the amplification of the small signal from the cartridge; and the power amplifier, which deals with the high-power amplification necessary to drive the speaker.

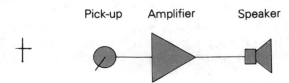

Figure 14.7 A block diagram for a record-player amplifier.

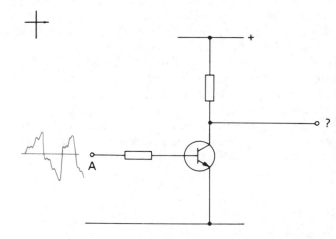

Figure 14.8 A simplified transistor amplifier. This circuit could not amplify an audio signal.

At its simplest a transistor amplifier circuit looks like the one in Figure 14.8, but although this simple configuration is satisfactory for demonstrating 'transistor action', it is incapable of amplifying an audio signal.

Consider the effect of applying the alternating voltage from the pick-up cartridge to terminal A as shown. Since the transistor conducts only when the base–emitter junction is forward-biased, only the parts of the signal that are positive relative to the emitter will cause the transistor to conduct. As well as amplifying the input signal, the transistor is rectifying it. Occasionally this happens in an amplifier under fault conditions, and it sounds horrible!

It is also clear that a proportion of the positive part of the signal is lost as well, since the transistor will not conduct, even with the base–emitter junction forward-biased, until the potential exceeds 0.7 V for a silicon transistor. This is a substantially higher voltage than a magnetic pick-up cartridge produces, so in practice the output of the circuit in Figure 14.8 would be nothing.

A solution to the problem is simply to connect a suitable positive potential to the base, ensuring that it is always forward-biased. This is best done with a potential divider network involving two resistors, as shown in Figure 14.9. This does at least produce an output, if the two bias resistors R_1 and R_2 are exactly the right values; the resistors should be chosen so that the collector is at half the line voltage with no signal applied. This gives approximately equal 'headroom' for the signal in its positive and negative excursions. Referring to the characteristic curves in Chapter 10 (Figure 10.12), this means that the operating point – that is, the mid-point of the variations in collector voltage when the amplifier is working – is halfway along the load line.

Because of tiny differences in manufacture it is not possible to specify the gain of a transistor very accurately; a gain variation of ±50 per cent is common. Using the circuit in Figure 14.9 would

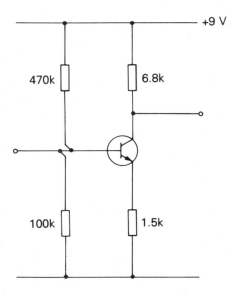

Figure 14.10 *An automatic bias circuit that will give correct operating conditions for transistors with a wide range of gains.

mean careful measurements and a new pair of resistors for each individual transistor used, which is unsuitable for mass production methods.

The leakage current of the transistor, and thus its operating point, will change with temperature. What is needed is a circuit that will automatically compensate for variations in gain of the transistor and ambient temperature. Such a circuit is shown in Figure 14.10. In this circuit it is the values of the resistors that determine the operating point. The gain of the transistor, leakage and temperature have practically no effect. The circuit involves a **feedback loop**. **Feedback** is a principle that is often used in electronics, and you will find it in many different contexts. In this case **negative feedback** is used. Negative feedback consists of a connection from the output of a system back to the input, arranged so that the change in output reduces whatever is causing that change. Negative feedback improves the stability of a system because any change is counteracted.

Although the circuit of Figure 14.10 looks simple, it is not immediately obvious how it works. The base of the transistor is held at about + 1.5 V, high enough to overcome the 0.7 V potential barrier of the transistor's base–emitter junction if the collector is near 0 V. Consider what would happen if the transistor were non-conducting; the emitter is connected to 0 V via the 10.5 kΩ resistor, and so is at 0 V. However, this means that there is a potential difference of about 1.5 V across the

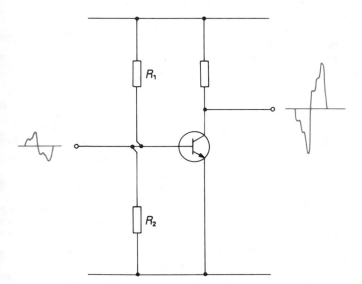

Figure 14.9 A bias network for a transistor, enabling the transistor to be used with audio signals. The resistor values would need to be altered for each individual transistor.

base–emitter junction, so the transistor would be turned on and would conduct.

This in turn means that a collector current is flowing; the collector resistor, transistor and emitter resistor operate as a potential divider. It is clear that the potential on the emitter will be higher than 0 V. Neglecting the voltage drop in the transistor, the emitter would be at a maximum potential of 1.5 V if the transistor were turned 'hard on'; but, of course, if the base and emitter are at roughly the same potential, there will be no base current and the transistor will be off!

In practice what happens is that the base current finds an equilibrium value, and in the circuit shown the collector will be 'balanced' at about half the supply voltage.

The compensating action of the circuit is easy to understand if you think what happens if conditions change. Consider what happens if an increase in temperature causes the leakage current to rise and the transistor's collector–emitter resistance to drop. The resistance in the upper half of the potential divider (of which the transistor's emitter is the middle) will be lowered, and the positive potential on the emitter will rise, to become closer to that of the base. This will reduce the base current and compensate for the temperature change. Transistors with different values of h_{FE} are also compensated for in the same way.

A simpler but slightly less effective bias circuit is shown in Figure 14.11. In this design the bias is fed to the base via a resistor connected between the collector and the base. Any tendency for the transistor to conduct more will reduce the positive potential at the collector, since the transistor and the 10 kΩ load (collector) resistor form the two halves of a potential divider. Ohm's law tells us that this reduces the current flow through the

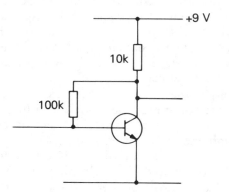

Figure 14.11 *A simpler circuit than that of Figure 14.10 is sufficient to provide automatic bias for silicon transistors in many applications.

100 kΩ bias resistor, because the voltage across it is reduced.

So this is another circuit in which any increase in the transistor's collector current causes a drop in base current.

Capacitor coupling

When an amplifier is used to amplify any continuously changing quantity (such as an audio signal), the amplifier shown in Figure 14.10 would be connected to the input device (a pick-up cartridge was mentioned) through a **capacitor**. The purpose of this is to block any d.c. component in the input. Figure 14.12 illustrates an amplifier circuit connected to a magnetic pick-up.

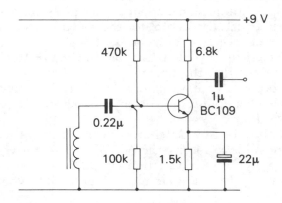

Figure 14.12 *A record-player pick-up preamplifier. Small capacitors couple the input and output of the amplifier, and a large electrolytic capacitor is used to bypass the emitter resistor, which improves the efficiency of the amplifier without altering the d.c. bias conditions.

The pick-up is a signal source, and this is **coupled** to the base of the transistor with a capacitor. The capacitor blocks any flow of d.c. that would otherwise interfere with the bias arrangements. You can see why this capacitor is needed by imagining what would happen without it. The resistance of the coil in the pick-up (about 30 Ω) would be in parallel with the lower resistor in the base bias system. This would reduce the potential on the base almost to 0 V, which would turn the transistor off. For the same reason, the output of the amplifier would be coupled to the next stage of amplification with a capacitor.

The emitter resistor reduces the efficiency of the amplifier by allowing the potential of the emitter to change with the amplitude of the signal being amplified. A large-value capacitor, connected across the emitter resistor, prevents the signal component affecting the bias conditions. An electrolytic capacitor is generally used for the emitter resistor **bypass capacitor**, so called because it bypasses the resistor at audio frequencies. Figure 14.13 illustrates a two-stage amplifier based on the circuit in Figure 14.12, with capacitor coupling between stages. Note the difference in resistor values for each stage: the second stage has a larger input voltage swing than the first, and also delivers a higher output current.

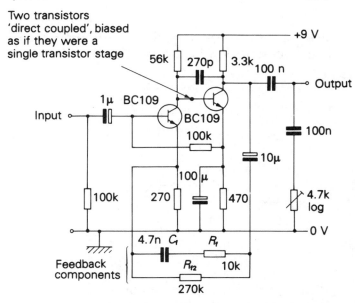

Figure 14.14 *A commercially produced amplifier, intended for use with a tape-player. The input is provided by a tape playback head, and the output is fed to an audio amplifier.

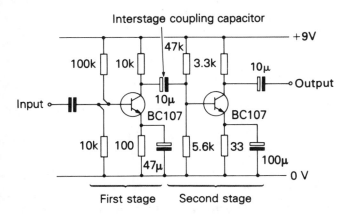

Figure 14.13 *A two-stage amplifier, using the same circuit configuration as that of Figure 14.12.

Gain of multistage amplifiers

Calculating the total gain of amplifiers having more than one stage is easy. All you have to remember is that the total gain of the system is equal to the gains of each individual stage multiplied together. If the gain of both stages in a two-stage amplifier were 30, then the total gain of the system would be 30 × 30 = 900.

Negative feedback

Negative feedback is often applied over more than one stage. Figure 14.14 shows a commercially produced design incorporating this feature.

It is possible to use feedback to alter the **frequency characteristics** of an amplifier. It is often necessary to limit the high-frequency amplification of a system, and this is easily done using a feedback capacitor instead of a resistor. The capacitor's use as a 'frequency-sensitive resistor' is clear. At low frequencies the small value of capacitance feeds

back only a tiny proportion of the signal, whereas a progressively larger proportion of the signal is fed back as the signal frequency increases.

In Figure 14.14 the feedback line is from the collector of the second transistor to the emitter of the first transistor. It is coupled via a large capacitor (10 μF) so that it does not affect the bias conditions. Frequency-selective components in the feedback line establish the frequency response of the amplifier: C_f, R_f and R_{f2} are calculated so that this amplifier gives the right frequency response for a tape playback head. The amplifier was designed for a high-quality tape-player.

The 270 pF capacitor between the two transistor collectors operates to limit the high-frequency response of the amplifier, and the variable resistor across the output, in series with a capacitor, trims the overall frequency to suit the amplifier that follows.

Transformer coupling

Transistor stages are generally coupled with capacitors, as in the preceding three circuits, but they can also be coupled with transformers. This used to be the preferred method, but as components and circuit design have improved, transformers (which are expensive and large) have been used less and less.

The resistance of the transformer's primary winding can form the collector load of the first stage of a two-stage amplifier. Transformer coupling is illustrated in Figure 14.15.

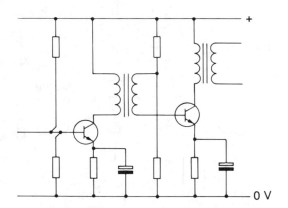

Figure 14.15 A two-stage amplifier using transformer coupling between stages and at the output. Transformer coupling is now seldom used.

Input impedance

The input impedance of an amplifier is the ratio of the input voltage to input current. It is expressed in ohms, and can be thought of as being similar to resistance, but is applicable to alternating voltages and currents. The input impedance of an amplifier is an important parameter, and is considered when the amplifier is connected to an external signal source, such as a record-player pick-up or tape head.

A cheap record-player might well use a crystal pick-up cartridge that generates a relatively high voltage output signal – as much as a volt – but has a very high internal resistance. The input circuit of an amplifier and pick-up cartridge is shown in Figure 14.16.

If the cartridge has an internal resistance of 5 MΩ – a typical value – and the amplifier has an input impedance of 2500 Ω, then the total current flowing in the input circuit (Figure 14.16) will be

$$I = \frac{V}{R}$$
$$I = \frac{1}{5\ 002\ 500}$$
$$I = 0.1\ \mu A$$

which means that of the total 1 V output of the cartridge, only 2500/5 000 000, about 0.5 mV (0.0005%) will be available to the amplifier. Not very efficient!

An amplifier with a higher input impedance should therefore be used to maximise the efficiency and make best use of the high output voltage of the crystal pick-up. Using a FET is a good idea

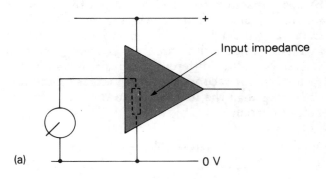

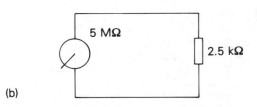

Figure 14.16 (a) A record-player pick-up, connected to an amplifier. The impedance 'seen' by the pick-up affects the efficiency with which power is transferred from the pick-up to the amplifier. (b) A high-impedance (crystal) pick-up feeding into a low-impedance amplifier is not very efficient.

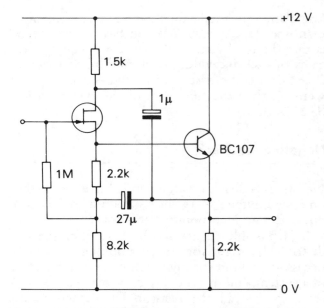

Figure 14.17 *A FET used to give a high input impedance.

because a FET has a very high input resistance. A commercial circuit using a JUGFET is given in Figure 14.17. The input impedance is about 5 MΩ.

Operational amplifiers and other d.c. amplifiers

You will quite often find transistor amplifiers, such as those described, used in commercially produced equipment, but there is a trend towards using **integrated circuits** because they tend to be cheaper, perform better, and require less work to assemble than amplifiers that use 'discrete' components. It is easy to see why. The amplifier shown in Figure 14.14 has 18 components and a minimum of about 40 soldered connections to mount them on a circuit board. An IC designed to do the same job could easily have hundreds of components, but would have less than half as many soldered connections.

A type of integrated circuit known as an **operational amplifier** ('op-amp' for short) is one of the most useful ICs, particularly for students! Op-amps are general-purpose d.c. amplifiers (the name dates back to the days of analogue computers) and are designed to be as close to a 'perfect' amplifier as possible.

> Although your syllabus may not ask you to study operational amplifiers specifically, the fact that an op-amp is really no more than a conveniently packaged d.c. amplifier means that the following section is equally applicable to 'discrete' amplifiers made up from separate transistors, resistors and capacitors.

The op-amp's circuit diagram is very similar to the general block diagram for an amplifier. It is shown in Figure 14.18.

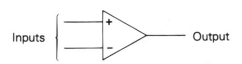

Figure 14.18 The circuit diagram for an operational amplifier.

Op-amps have one output connection, but two input connections, called the **inverting** and **non-inverting** inputs.

The specification of a 'perfect' operational amplifier might look like this:

1. **Gain.** This should ideally be infinitely high. Obviously an amplifier with infinite gain would be useless, as the smallest input would result in full output! A very high or even infinite gain can be controlled by suitable feedback.
2. **Input resistance.** Ideally, this should also be infinite, so that there is no loading of the input source at all.
3. **Output resistance.** Ideally, this should be zero. With a zero output resistance, the amplifier can be connected to a load of any resistance without the output voltage being affected.
4. **Common mode rejection ratio.** An operational amplifier has one output, but two inputs, an inverting input and a non-inverting input. Figure 14.18 shows an op-amp system, with the two inputs clearly marked: '−' for inverting and '+' for non-inverting. A positive voltage applied to the inverting input makes the output swing negative and a positive voltage applied to the non-inverting input makes the output swing positive. It is most important to understand that the inputs are relative to *each other* and *not* to

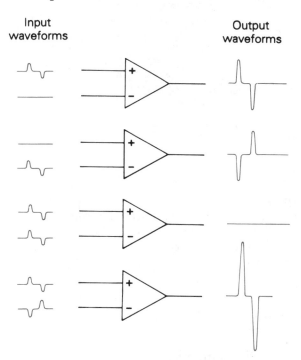

Figure 14.19 Showing how an op-amp responds to different kinds of input.

either of the supply lines. Thus, if both inputs (common mode) are made more positive or negative, there should be no output. This is illustrated in Figure 14.19.

5. **Supply voltage.** The amplifier should be unaffected by reasonable variations in its power supply voltage.

Now we can compare the theoretical specifications with a real one: the specification for the SN72741 op-amp.

1. Gain: 200 000 V voltage gain.
2. Output resistance: 75 Ω.
3. Input resistance: 2 MΩ.
4. Common mode rejection ratio: a signal applied to both inputs will be at least 32 000 times smaller than the same signal applied to just one input.
5. Supply voltage: output will change less than 150 µV per volt change in the power supply. The amplifier will operate from a supply of ±3–18 V (see below) and uses about 2 mA when there is no input. The maximum power dissipation is 50 mW, and the maximum output current is around 30 mA.

Figure 14.20 shows the connections to this integrated circuit, which is most commonly available in an 8-pin DIL pack.

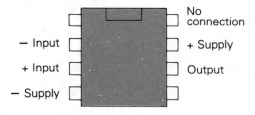

Figure 14.20 The SN72741 op-amp integrated circuit.

In looking at the ways in which op-amps are used in various configurations, remember that you can substitute a 'discrete' d.c. amplifier in any of these circuits; you can make the same calculations, and get the same results!

Negative feedback

An inverting amplifier
First of all we need some control over all that gain! Figure 14.21 illustrates the basic op-amp configur-

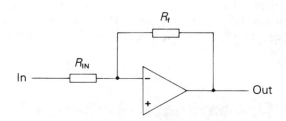

Figure 14.21 Basic op-amp negative feedback configuration, using the inverting input.

ation with negative feedback. If the amplifier has a very large (assumed to be infinite) gain, the actual gain is controlled only by the values of the resistors R_{IN} and R_f. R_{IN} is the input resistor, and must be substantially less than the input resistance of the op-amp. R_f is the feedback resistor. The voltage gain of the system is very simply calculated as

$$A = \frac{R_f}{R_{IN}}$$

where A is the voltage gain.

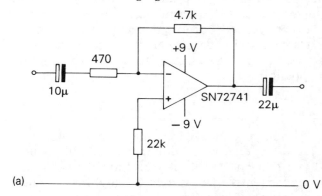

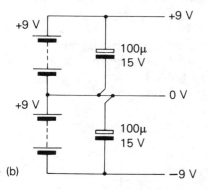

Figure 14.22 ★(a) A practical amplifier based on the circuit of Figure 14.21. ★(b) A simple but suitable power supply.

The fact that the op-amp has finite gain affects the calculation only slightly and is generally ignored (this is not necessarily so with a discrete amplifier), provided the required gain is not approaching the specified maximum.

Figure 14.22(a) is the practical circuit. Notice the odd power-supply requirements. The SN72741 needs a power supply that is symmetrical about zero. This permits the output to swing above and below zero. There are various ways of contriving such a supply, but the simplest (and good enough for our purposes) is to use 9 V batteries (PP3 or PP9) wired as shown in Figure 14.22(b).

Remember that the inverting input is relative to the non-inverting input, not to the 0 V supply line. In our simple amplifier we want the input to be relative to 0 V ('earth'), so we connect the non-inverting input to 0 V. This is best done via the resistor, though the value is uncritical, and 22 kΩ is convenient. This amplifier circuit has a gain of 10, set by the values of the input and feedback resistors, 4700/470. The capacitors are added for a.c. operation (see above).

A non-inverting amplifier

The amplifier shown **inverts** the input. The design of a **non-inverting** amplifier is slightly more difficult, since although the input is applied to the non-inverting input, the feedback still has to be applied to the inverting input. The basic configuration is shown in Figure 14.23.

A practical non-inverting amplifier with a gain of 11 is shown in Figure 14.24.

It is possible to apply several inputs to the op-amp configuration, isolating the inputs from one to another with the input resistors, which should be as high in value as practicable: the higher they are, the better the isolation. Such a circuit could be used for an audio mixer, as shown in Figure 14.25. The input resistors R_1, R_2 and R_3 set the maximum gain of the channels; in this case, R_1 and R_2 give a gain of 10, and R_3 gives a gain of unity (no amplification). The logarithmic potentiometers provide volume controls that are suitable for audio use: slider controls are more convenient to use than rotary ones.

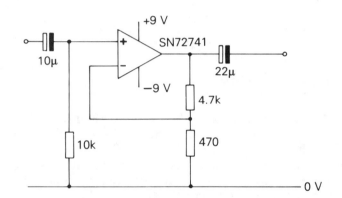

Figure 14.24 *A practical circuit based on the configuration in Figure 14.23. The power supply in Figure 14.22(b) can be used for this circuit.

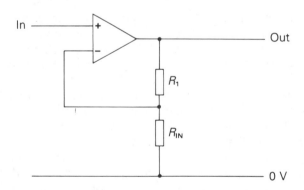

Figure 14.23 The basic non-inverting configuration.

The gain of the non-inverting amplifier shown here is calculated as

$$A = \frac{R_{IN} + R_f}{R_{IN}}$$

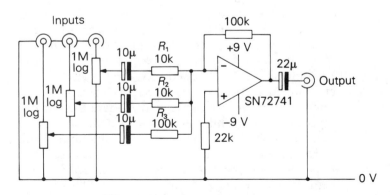

Figure 14.25 *A practical circuit for an audio mixer; an op-amp provides amplification. The power-supply circuit of Figure 14.22 (b) can be used.

Audio power amplifiers

The decibel

Power amplifiers are used to provide a high-power output – in this case at audio frequencies – to drive a speaker with sufficient volume to hear clearly. The **power gain** of such an amplifier is usually measured in **decibels**. A decibel is one tenth of a **bel** (B), which is equal to the logarithm of the output power (P_{out}) divided by the input power (P_{in}):

$$B = \log \left(\frac{P_{out}}{P_{in}} \right)$$

so

$$dB = 10 \times \log \left(\frac{P_{out}}{P_{in}} \right)$$

Class A amplifiers

Audio-frequency amplifiers (amplifiers designed to amplify signals at around the frequencies to which the human ear is sensitive) have their own special problems, and the circuits are designed to solve these problems. The range of frequencies to be amplified is well within the capability of even simple circuits. The generally accepted 'hi-fi' frequency range is between 20 Hz and 20 kHz, although only young children can hear frequencies as high as 20 kHz, and at frequencies as low as 20 Hz the sound is felt rather than heard.

The first requirement for an audio amplifier to be considered here is the power output. A typical transistor radio or portable tape-player will provide an output power of around 500 mW into a small speaker. Personal stereo systems like the Walkman® are designed for use with high-efficiency earphones and have an output power of only a few milliwatts. For 'hi-fi', where the faithful reproduction of high-energy transient sounds is important, 20 W is considered a sensible minimum.

To deliver about 500 mW into the speaker requires about 55 mA from a 9 V supply, if we neglect any inefficiencies in the system. This power output will be required only rarely: the loudest sounds, with the volume turned well up. This appears to be well within the capabilities of a handful of small dry batteries.

You will (I hope) recall from the beginning of this chapter that in order to amplify audio signals in an undistorted way, the output of an amplifier has to be able to swing positive and negative of a neutral point. This is achieved in amplifiers like the one in Figure 14.26 by biasing the transistor so that the collector is at approximately half the supply voltage.

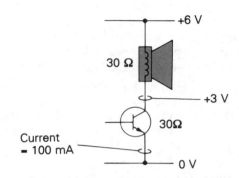

Figure 14.26 A simple way of driving a speaker with a transistor amplifier.

In the simple amplifier in Figure 14.26, this happens when the resistance across the transistor's emitter–collector junction is equal to the resistance of the collector load; the transistor and the load resistor are then the two halves of a potential divider with equal resistance either side, and the mid-point will be at half the supply voltage. A speaker typically has a resistance of 30 Ω, so the total resistance is about 60 Ω. Thus the current taken by this output is roughly 6 × 1000/6 = 100 mA. For small and medium-sized dry batteries, this represents a fairly short life. The current is consumed even when there is no output from the speaker. In fact, the current taken from the supply will, on average, always be the same, since for the most part audio signals will reduce and then increase the transistor's emitter–collector resistance symmetrically.

There is also the question of what happens to the power lost from the batteries. Half is dissipated by the transistor, and the other half by the speaker: 300 mW each. Thus cooling, even in a modest amplifier, is something that needs to be taken into consideration.

And finally there is the problem caused by 100 mA flowing continuously through the speaker, pulling the speaker cone out of its true central position. It means that a small speaker could not be used, and for this (and other) considerations, the whole amplifier is going to be quite large.

Although this design seems hopelessly unsuitable, a complete amplifier based on the simple circuit above is shown in Figure 14.27. This was produced commercially in the 1960s, and actually makes a good, musical sound. It would be quite an interesting project to make it, as the transistors (or their equivalents – the circuit is pretty tolerant) are still available.

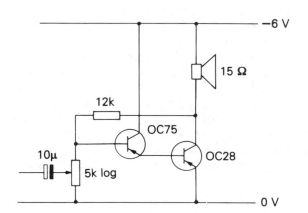

Figure 14.27 *A practical direct-coupled Class A audio amplifier.

The transistor will need a heat sink of some sort (a piece of aluminium 1 mm thick by 100 mm square would do), and a speaker of at least 150 mm diameter is recommended. Speakers with 30 Ω coils could be hard to find, so the circuit values have been changed for a 15 Ω speaker. The circuit takes about 300 mA from a 6 V battery. The circuit is designed for a crystal record-player cartridge (they were widespread in the 1960s and have a high output). This amplifier hasn't a great deal of gain.

The amplifier we have been discussing is known as a Class A audio amplifier. A properly designed Class A amplifier can have very low distortion, but will always waste a great deal of power.

Class B amplifiers

Today all audio power amplifiers are Class B amplifiers. The main reason is that a Class B amplifier consumes power that is much nearer to being proportional to the output power at any instant. In other words, a loud signal will cause the amplifier to draw a heavy current from the supply, whereas when quiescent (no signal) it will take very little current. Look at Figure 14.28.

The circuit depends for its operation on the fact that the transistors require different polarity of signal to drive them. The input waveform is shown in the diagram: one cycle of a sine wave at an audio

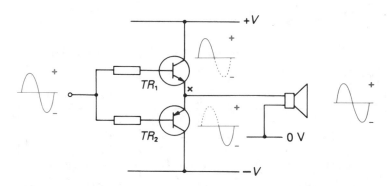

Figure 14.28 One of the basic forms of the Class B amplifier. This is known as a complementary-symmetry Class B amplifier.

frequency. The first part of the signal, the positive half-cycle, makes TR_1 conduct and act as an amplifier. TR_2, however, will be turned off (non-conducting) by the signal. This means that point X on the diagram, the output to the speaker, is the output of the transistor amplifier TR_1. The speaker's speech coil is the emitter load of this amplifier.

The second (negative) half-cycle of the input signal switches TR_1 off, but makes TR_2 conduct. This time TR_2 is the amplifier, with the speaker still the emitter load of this amplifier. The result is that the two halves of the input signal are handled by two different transistors, but the output from both transistors goes through the speaker, to reconstitute the original signal. The efficiency of the amplifier comes from the fact that, with no input signal, neither transistor is conducting, so no power is taken from the supply.

Like most things, it turns out that the real circuit isn't quite so simple. Transistors do not respond to a signal in a linear manner when operated near cut-off point (look at the characteristics graphs in Chapter 10, Figure 10.12). The transistors have to be biased so that they are just conducting enough to move them into the linear part of the characteristic. Distortion will then be minimised.

Distortion at the point where one transistor takes over from the other is that most often encountered, especially in cheap or poorly adjusted designs. Such distortion is called **crossover distortion**. An example of crossover distortion in a poorly adjusted design is shown in Figure 14.29. The input is a sine wave.

A circuit with a suitable bias system to maintain the transistors at the correct point on the curve is illustrated in Figure 14.30. The diodes have a secondary purpose, that of providing the amplifier with thermal stabilisation. For the lowest possible

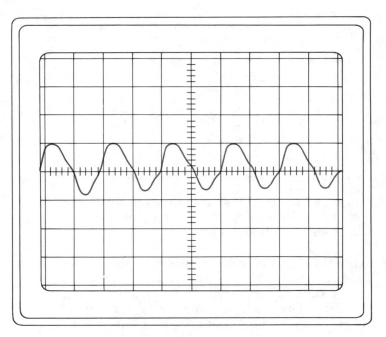

Figure 14.29 Crossover distortion: amplifier output, with a sine-wave input.

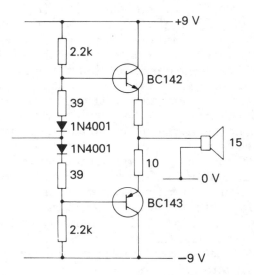

Figure 14.30 *A practical Class B output stage to drive a 15 Ω speaker. The diodes are in contact with the encapsulations of the transistors and provide thermal compensation.

crossover distortion, the operating points of the transistors need to be set very accurately, and when the amplifier is working – particularly if it is delivering a large output – the transistors will get hot, changing the characteristic curves and necessitating a slightly different bias setting.

Automatic compensation is obtained by putting

the diodes in contact with the transistors, so that the increase in temperature of the transistor also heats the diode. The diode's conduction changes correspondingly with temperature, altering the bias conditions so as to compensate (approximately) for the transistors' changes. The system is surprisingly effective.

The amplifier in Figure 14.30 uses a split supply, like the op-amps. This is an efficient mode of working, but the use of a split supply may well be inconvenient in a small portable device. It is possible to use a single supply by placing a capacitor in series with the speaker to prevent a direct current flowing all the time. The layout is shown in Figure 14.31. Because of the large currents flowing, the capacitor must be very large: in the order of hundreds of microfarads. Often it will be by far the largest physical component in the circuit.

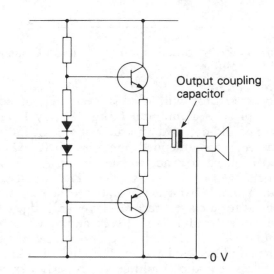

Figure 14.31 The requirement for a split power supply can be obviated by the use of an output-coupling capacitor.

In the design of the Class B output stage we have concentrated on power-handling and efficiency. Gain is a low priority, and the amplifier will invariably use additional low-power stages to amplify the incoming signal to a level where it can drive the output transistors.

Before leaving the subject of Class B power output stages, it is important to realise that the complementary-symmetry design is by no means the only one possible. It is the one that is most often used because it is simple and works well. It is possible to make an amplifier that has transformer coupling of the input and output stages, and identical transistors on both sides of what is sometimes called a 'push–pull' Class B layout. The transformer coupling to the input is used to provide the two halves of the system with input signals of opposite phase. An output transformer may be used to match the speaker impedance to the output stage more closely. A transformer-coupled design is shown in Figure 14.32.

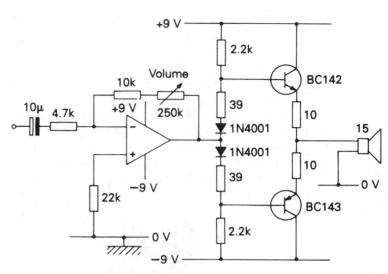

Figure 14.33 *An op-amp preamplifier stage, connected to the power amplifier.

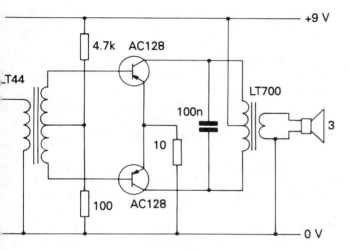

Figure 14.32 A transformer-coupled Class B amplifier.

Transformer coupling is now used rarely, since the price of transformers and their size make them undesirable for modern circuits. The transformer also adversely affects the frequency response and distortion.

Preamplifier and driver stages

Often the Class B output stage is preceded by a single transistor amplifier that provides some of the gain needed for the amplifier. This in turn is preceded by a preamplifier, which provides most of the gain and may also incorporate tone and volume controls. Any division between the driver and preamplifier is rather artificial, and today the whole system coming before the power amplifier tends to be termed 'preamplifier'.

A simple preamplifier can be made with the ubiquitous op-amp. The circuit is given in Figure 14.33. The volume control is in the feedback line and works by altering the overall gain of the preamplifier. This entirely practical circuit has quite a respectable performance. The frequency response should be only about 3 dB down at 20 Hz and 25 kHz; it has an input impedance of 10 kΩ, and will deliver about 200 mW into a 15 Ω speaker.

Integrated circuit audio amplifiers

The trend in electronics today is towards fewer and fewer components in a given system. An audio amplifier is just the sort of device that is suitable for production as an integrated circuit: there is a large market for identical units, and all components in the system (most of them, anyway) can be produced using electronic techniques. It is not surprising, therefore, that there is no justification nowadays for building an audio amplifier from discrete components, unless it is for hi-fi applications.

An example of just what can be done in integrated circuit audio amplifiers is the LM380, a 2 W audio amplifier. Including the volume control, the complete amplifier uses six components. The full circuit is shown in Figure 14.34. The LM380 has obvious similarities to the op-amp; there are two inputs, inverting and non-inverting, but it has a

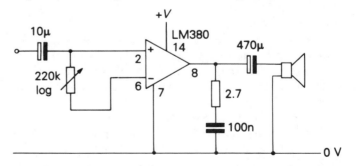

Figure 14.34 *A modern integrated circuit audio amplifier.

Class B output stage. It will deliver 2 W maximum into an 8 Ω speaker when connected to a 20 V supply, but will work with any supply voltage down to a few volts.

Figure 14.35 shows the pin connection; the circuit is in the familiar 14-pin DIL pack. The middle six pins are connected to the 0 V supply line, but also provide a heat-sink path for the output transistors. If the printed circuit board is designed with a large area of copper (say, 20 cm²) connected to these pins, then heat is conducted away from the output transistors and is dissipated by the copper of the printed board itself: a neat design idea!

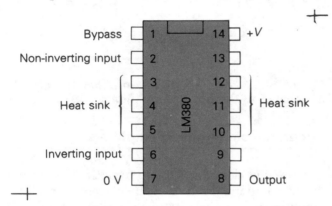

Figure 14.35 The pin connections of the LM380 audio amplifier circuit. The centre three pins on each side are connected to the output transistors and provide a heat-sink path.

■ CHECK YOUR UNDERSTANDING

● **Power supply** circuits convert mains electricity into the direct or alternating voltages required by an item of equipment. A simple power supply circuit consists of a **transformer** to reduce the mains voltage, a **rectifier** to change a.c. to d.c., a **smoothing capacitor** to smooth out short-term variations in the direct voltage, and (if necessary) a **regulator** to ensure that the direct voltage remains constant at the right level.

● Audio amplifiers are usually divided into two sections: the **preamplifier**, which incorporates the tone and volume controls; and the **power amplifier**, which provides a powerful signal to drive the speakers.

● Transistors – when used as amplifiers – need **bias resistors** to turn the transistor on and make sure that it is working on the straight part of the characteristic curve. In a common-emitter amplifier the bias is set to give about half the supply voltage at the collector, with no input signal.

● Bias resistors are invariably connected to make an **automatic** bias circuit that compensates for differences in gain between one transistor and another, and the effects of changing temperature.

● Audio amplifier transistor stages are usually connected together by capacitors (**capacitor coupling**). These block direct currents and prevent a transistor's biasing from being affected by the preceding or following stage. The inputs to audio amplifiers are connected to the signal source through capacitors for the same reason.

● **Negative feedback** is used to control the gain and frequency response of an amplifier.

● The **input impedance** of an amplifier must be matched to the impedance of the signal source. The **output impedance** must be matched to that of the output device: in audio amplifiers, this is a speaker.

● **Operational amplifiers** are integrated circuit d.c. amplifiers that are designed to match the properties of an 'ideal' amplifier as closely as possible.

● **Power amplifiers** are usually **Class A**, which can be made to give a good audio performance but waste power and have to be designed to dissipate a lot of heat, and **Class B**, which are more difficult to design but use an amount of power that is roughly proportional to the loudness of the output.

● All except some of the most expensive audio amplifiers are built as **integrated circuits** to save money and improve reliability.

REVISION EXERCISES AND QUESTIONS

1 Draw a circuit diagram (you do not have to put in the component values) for a power supply involving a transformer, bridge rectifier, and smoothing capacitor.
2 Draw a circuit diagram for a Zener diode regulator that might follow the circuit shown in (1) above.
3 Sketch a simple one-transistor amplifier with an automatic bias circuit.
4 Why is **capacitor coupling** often used between stages of a multistage amplifier?
5 Sketch the basic op-amp configuration using the inverting input for negative feedback.
6 What is the main disadvantage of a Class A power amplifier?
7 Class B amplifiers overcome the problem in (6) above. How?
8 What is meant by 'input impedance'?

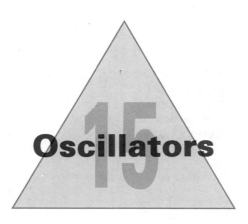

Oscillators

Introduction

It is often necessary to generate a continuously changing voltage or current. The frequency of change of an oscillator is designed according to the applications, but can be anything from one cycle in several hours to hundreds of megahertz. All circuits that generate a simple steady-state repetitive signal are called **oscillators**.

We saw how a **unijunction oscillator** works in Chapter 12. The unijunction oscillator is a form of **relaxation oscillator**, but there are many other types.

Relaxation oscillators

Gas-discharge relaxation oscillator

About the simplest form of oscillator is the relaxation oscillator, shown in the circuit in Figure 15.1. It derives its name from the fact that the circuit alternately conducts and 'relaxes'. This rather old-fashioned circuit makes a good demonstration because you can see it working! It uses a neon lamp. The neon lamp is filled with neon gas at low pressure. When a sufficiently high voltage is applied to the lamp terminals, the neon gas begins to conduct electricity, and at the same time glows red. The lamp will continue to conduct (and glow) until the voltage drops to a level rather lower than that required to 'strike' the neon. This is important: the voltage required to initiate conduction is higher than the voltage required to keep the lamp on.

When the circuit is connected to a voltage source shown (a high-voltage battery, for example) the

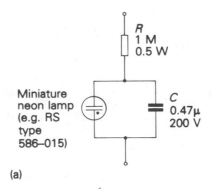

(a)

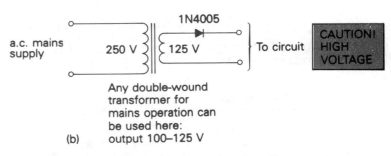

(b)

Figure 15.1 *(a) A simple relaxation oscillator, using a neon lamp and capacitor. *(b) A suitable power supply for the circuit in Figure 15.1(a). The diode must have a working voltage of at least 200 V.

capacitor C is uncharged. It slowly charges up via resistor R, and the voltage across it increases. At the point where the voltage across the capacitor (and, of course across the neon lamp terminals) reaches the striking voltage of the neon, the lamp abruptly turns on.

Current from the capacitor now flows through the lamp, which rapidly discharges the capacitor until the voltage across it drops below the level required to sustain the lamp. The lamp goes out and the cycle repeats.

The circuit shown in Figure 15.1(a) is useful for demonstrations, but requires a d.c. power source of about 150 V. A suitable power supply is shown in Figure 15.1(b). For safety this circuit must not be used direct from a.c. mains. The output waveform of this circuit is illustrated in Figure 15.2.

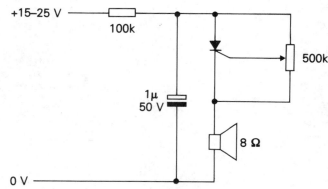

Figure 15.3 *A thyristor used in a relaxation oscillator.

ible 'pop'. Once the capacitor is fully discharged there is insufficient current available to keep it switched on (for an 18 V supply, 18/100 mA = 180 µA). The thyristor begins to charge up again, and the cycle is repeated. The potentiometer (and of course the supply voltage) control the frequency of oscillation. Try the effect of different component values.

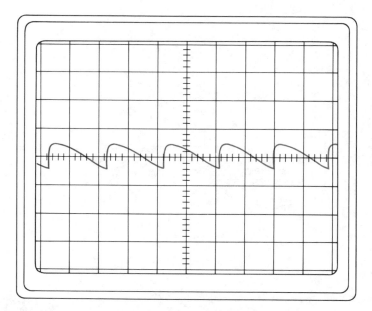

Figure 15.2 The output of the gas-discharge relaxation oscillator.

LC oscillators

We begin by looking at the behaviour of a **tuned circuit**. A capacitor (symbol C) and inductor (symbol L) are connected in parallel (Figure 15.4).

Thyristor relaxation oscillator

A thyristor can be used to make a relaxation oscillator. This makes a more interesting demonstration than the neon oscillator because it produces an audible output. It also operates from a somewhat lower voltage supply: two 9 V dry batteries in series can be used as a power source. The circuit is given in Figure 15.3.

It is easy to see how this circuit works. The 1 µF capacitor charges through the 100 kΩ resistor and the voltage across it gradually rises. Because the speaker has a very low resistance, almost the whole of this voltage appears across the 500 kΩ potentiometer. If the brush of the potentiometer is half-way along its travel (for instance), half this voltage will appear across the gate-cathode junction of the thyristor. When it rises to a high enough level the thyristor will be triggered.

The 1 µF capacitor now discharges rapidly through the thyristor and speaker, causing an aud-

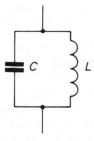

Figure 15.4 A simple *LC* tuned circuit.

If a direct voltage power supply is connected to this circuit, the capacitor will charge up, but little current will flow initially through the inductor (re-read the section on Inductors in Chapter 2 if you can't see why). If the power supply is immediately disconnected, energy is left stored in the charged capacitor.

The capacitor now discharges into the inductor,

and the energy stored in the capacitor is converted into the magnetic field associated with the inductor. After a time, the capacitor will be discharged, and the current flow will stop. But with no current flow to sustain it, the magnetic field will begin to collapse, converting its energy back into electricity and recharging the capacitor! While this is happening, current flows round the circuit in the opposite direction.

Once the capacitor is charged again, the circuit is back in its original state, except that a small amount of energy will have been lost, ultimately radiated by the inductor as a tiny amount of heat.

This cycle of charge–discharge repeats continuously until all the energy has been lost. Figure 15.5 shows a graph of current flowing through the circuit.

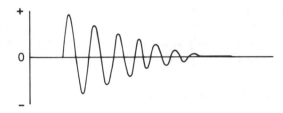

Figure 15.5 The p.d. across the circuit after it is energised. The amplitude gradually reduces to nothing as energy is lost, but the frequency remains constant.

> Notice that the frequency of oscillation is constant.

Every combination of inductor and capacitor has its own **resonant frequency**, in the same way that a clock pendulum has a resonant frequency. It is hard to describe resonance without resorting to mathematics that may be beyond the level of this book. I am therefore going to skip over a detailed description, but simply tell you that a parallel combination of inductor and capacitor appears, **at the resonant frequency only**, to have a much higher impedance (impedance is like resistance, only for alternating voltages) than at any other frequency.

The resonant frequency (f) in hertz of the LC oscillator can be calculated as follows:

$$f = \frac{1}{2\pi\sqrt{LC}}$$

It is possible to use the **resonance** of LC circuits

to make an **oscillator** that has a very stable frequency, and a sine-wave output.

The requirement for oscillation is some form of **positive feedback**, applied at the resonant frequency by the LC circuit. This is rather like giving a pendulum a tiny push at the end of each swing; it keeps it going, but without affecting the time it takes to swing back and forth.

Several designs are routinely used to achieve this. Figure 15.6 shows the simplest of these: the **Hartley oscillator**. The 100 nF capacitor provides the necessary feedback, while L and C control the frequency. This circuit needs an inductor that has a tapped winding: a connection made to the coil in the centre as well as to the ends. Otherwise, it is a simple and reliable circuit. The values shown on the circuit are suitable for an audio oscillator: L and C should be chosen to give a frequency of 1–10 kHz, if you wish to make up a demonstration circuit.

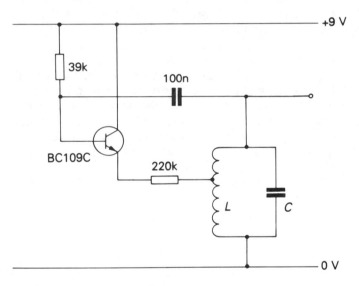

Figure 15.6 A Hartley oscillator.

LC oscillators have a stable frequency, and can easily be made to work at **radio frequencies**. More about radio in Chapter 16.

Crystal-controlled oscillators

Although LC oscillators can be made quite stable, the frequency of operation is affected to some extent by temperature and voltage fluctuations.

Both the capacitor's capacitance and the inductor's inductance are altered by temperature changes.

The search for a really stable oscillator led to the development of the quartz crystal oscillator. Central to this is a specialised component, the **quartz crystal**. When a crystal of quartz is subjected to an electric voltage, it flexes (we looked at this in Chapter 2: the **piezoelectric** effect). Conversely, when you bend the crystal, a small electrical voltage is generated across the crystal.

Quartz makes a very suitable material for an oscillator crystal because it is elastic; just like a pendulum, it takes a long time to stop oscillating once it has started. When 'ringing' at its resonant frequency, the quartz crystal is very like the *LC* circuit. Energy goes in, which flexes the crystal; the crystal 'swings' back, and electrical energy is produced; as it 'swings' back the other way, the voltage is reversed.

A typical crystal oscillator is illustrated in Figure 15.7. This one works at radio frequencies, in the HF band (see Chapter 16). The crystal itself provides the necessary feedback between the collector and base of the transistor. Note the use of the inductor in the collector line. This has a low d.c. resistance for the correct biasing levels, but has a high resistance at radio frequencies and allows the output to be taken between the collector and positive supply line. The circuit can be made to oscillate at a frequency determined *only* by a crystal, regardless of the precise values of other circuit components.

Quartz crystals can be made cheaply and accur-ately, and in suitable circuits can give astoundingly accurate control. This type of oscillator is used in digital watches, and an accuracy of ten seconds a month (within the reach of the cheapest digital watch) implies a long-term oscillator stability of better than 1 part in a $\frac{1}{4}$ million. Quartz crystals can be made to work over a range of frequencies from tens of kHz to a few MHz.

Crystal-controlled oscillators are often used in radio transmitters.

Transistor multivibrators

We shall next look at three related circuit types; only one of them is an oscillator, but it is convenient to consider them all together because they are so similar. The first of them is the **bistable multivibrator**, which isn't an oscillator, but is the best one with which to begin.

Bistable multivibrators

A circuit that can assume two stable states is known as a **bistable circuit**. Such a circuit is illustrated in Figure 15.8. Essentially, it consists of two simple transistor amplifier circuits, connected so that each transistor's base is connected, through a resistor, to the collector of the other.

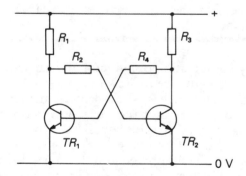

Figure 15.8 *A bistable multivibrator circuit.

If TR_1 is conducting, its collector will be only 0.7 V or so above 0 V, and TR_2 is therefore switched off; it is in the non-conducting state. TR_2, in contrast, has its collector at a voltage approaching that of the supply. TR_1 base is therefore held at a high potential, which keeps it on. The circuit is clearly stable in this configuration, and remains

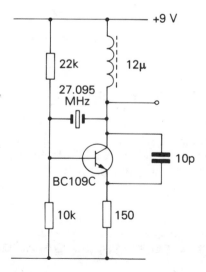

Figure 15.7 *A quartz-crystal-controlled oscillator. This provides extremely good frequency stability.

with TR_1 on and TR_2 off indefinitely. Because the circuit is symmetrical, it can equally well be stable with TR_1 off and TR_2 on.

Bistable circuits like this form the basis of many sorts of digital memory element. It can be 'forced' into one state or the other by momentarily connecting either T_1 or T_2 to the positive supply. The circuit will remain in either state as long as the power supply remains connected. Two LEDs can be inserted in series with R_1 and R_3. They don't affect the way the circuit works but indicate which side is 'on'.

Astable multivibrators

Figure 15.9 shows an astable multivibrator circuit. Assume TR_1 is on; it can only hold the base of TR_2 at a low potential until C_1 is fully charged. R_{B2} now turns TR_2 on, and the voltage applied to the base of TR_1 drops. To understand this circuit, it helps to consider R_{B1}, C_2 and the collector–emitter junction of TR_2, and to think about the changes in potential of the various points in the circuit when the capacitor is charged and uncharged, and TR_2 is on or off.

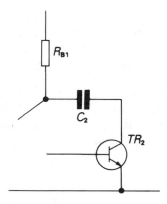

Figure 15.10 One 'leg' of the astable multivibrator.

impedance earphone, or even a small speaker in series with a resistor, will soon show it working.

Figure 15.11 gives the output waveform of the astable multivibrator circuit. For obvious reasons, this kind of waveform is called a 'square wave'.

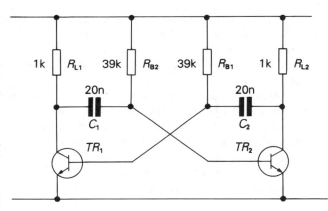

Figure 15.9 ★An astable multivibrator.

Look at Figure 15.10. TR_1 turns off and the circuit assumes the second 'stable' state, but only until C_2 charges up. Then the circuit swaps over again, TR_1 coming on and TR_2 turning off. This circuit will continue to switch back and forth between the two states at a rate that depends on the circuit values.

The lowest possible frequency of oscillation is generally limited by the size and leakage current of the capacitor; the highest frequency is limited by the type of transistor used. The circuit shown oscillates at around 1 kHz; an oscilloscope, a high-

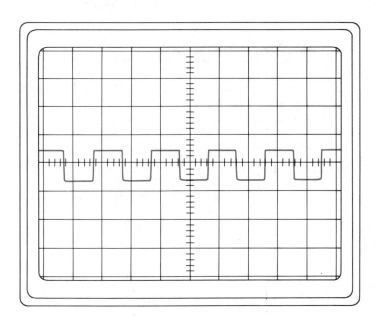

Figure 15.11 The output of the astable multivibrator.

Monostable multivibrators

The bistable circuit has two stable states, and the astable circuit has no stable state. I suppose it's reasonable to expect the monostable circuit to have one stable state, which in fact it does.

The circuit is given in Figure 15.12. It is – as you can see – a combination of the other two. It is stable in one state only. If T_1 is momentarily connected to 0 V, the circuit will change state, but will remain like it only for a time determined by the values of R and C. Study the circuit in the light of what you know about the functioning of the bistable and astable circuits and you should easily be able to work out how it operates.

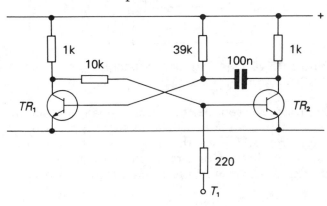

Figure 15.12 ★A monostable multivibrator.

Operational amplifiers as oscillators

Square-wave or sine-wave oscillators can be made with operational amplifiers. Figure 15.13 shows a basic op-amp oscillator.

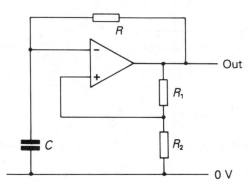

Figure 15.13 ★An op-amp oscillator.

The principle is straightforward. Starting with the amplifier in a condition where the output is positive, the capacitor C is charged, via R_1 and R_2, until the inverting input becomes sufficiently positive to cause the op-amp 'bistable' to change state: the output now becomes negative. It is now a negative potential that is applied to C, via R_1 and R_2. C is discharged and recharged with the other

polarity. When the potential on the inverting input changes, the 'bistable' again changes state; and so on.

The calculation for the rate at which the circuit changes state is given by the following formula, where $R = R_1 + R_2$ and T is the total time taken for the oscillator to go through one complete cycle:

$$T = 2RC \ln \left(1 + \frac{2R_1}{R_2}\right)$$

(ln is the natural logarithm, log to the base e).

■ CHECK YOUR UNDERSTANDING

● An **oscillator** is a circuit that generates a repetitive continuously changing signal.
● **Relaxation** oscillators are among the simplest.
● *LC* (inductor, capacitor) oscillators can produce a sine-wave output. They are stable and can be made to work at high frequencies.
● *LC* oscillators are used widely in radio receivers.
● **Crystal-controlled** oscillators are extremely stable and are used where a very stable and accurate frequency of oscillation is needed: for example, in clocks and watches, computers and radio transmitters.
● **Bistable multivibrators** have two stable states, and will remain indefinitely in either state.
● **Astable multivibrators** have no stable state: they are oscillators that produce a square-wave output.
● **Monostable multivibrators** have one stable state. If triggered by an input pulse they will temporarily change to the unstable state for a period determined by the component values, then change back again.
● Operational amplifiers can be used as oscillators when connected in suitable circuits.

REVISION EXERCISES AND QUESTIONS

1 What is an oscillator?
2 **Tuned circuits** using an inductor and a capacitor are often used where a stable sine-wave oscillator is needed. What is the formula used to calculate the resonant frequency of a parallel *LC* combination?
3 Sketch the circuit of a **bistable multivibrator**. How do you make it change state?
4 What kind of output waveform is produced by an **astable multivibrator**?

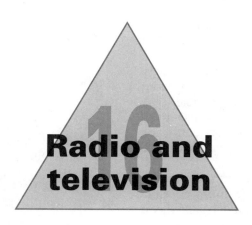

Radio and television

Introduction

Radio and television have brought the world closer together, allowing people in all countries to see how others live. If electronics consisted *only* of radio and television, it would still have had a tremendous impact on world culture, perhaps more than any other human invention since agriculture and the gun. Radio and television are very big subjects, and this chapter offers a very short introduction. It is intended to explain the basic principles, and to touch on the way that radio and television systems work. Because electronics *isn't* limited to radio and television, there is not enough space to go into more detail.

Radio waves and propagation

Radio transmission and reception was perhaps one of the earliest applications of electronics, and is – so far – the application that has made the greatest impact on society. We can use radio energy, predict its properties and design circuits that work efficiently, but in truth we know little about the real nature of radio.

Ask an electronics engineer what radio is, and the answer will be a confident 'electromagnetic waves'. Ask a physicist what electromagnetic waves are, and he will begin to hedge, or he will tell you that really we don't know. We *do* know that electromagnetic radiation is a form of energy, and that it behaves as if it is propagated as waves.

You may remember what I said in Chapter 2 about not confusing the model with the reality. This particular model looks less like reality when we consider that radio travels through a vacuum. How can there be waves in a vacuum? Perhaps in

the future, theoretical physics will give us an answer. In the meantime, we *use* radio, *describe* it mathematically, and design and use electronic circuits that function despite our underlying ignorance.

Radio transmitters

Possibly the hardest concept to grasp is the way in which circuits can be made that broadcast radio waves into their surroundings.

In Chapter 15, we looked at the operation of the *LC* oscillator. This (you will no doubt recall) is a form of oscillator that produces a stable sine-wave output, with a capacitor (C) and an inductor (L) controlling the frequency. It is particularly suitable for use at high (radio) frequencies. You may think it worth while re-reading the section.

All capacitors that we use in electronic circuits develop an **electric field** between two parallel and very closely spaced plates. The energy that is stored in the capacitor is in the electric field (whatever *that* may be: it depends on which model we are using!). Suppose we move the plates further apart. This has two effects. First, it decreases the capacitance (this principle is used in variable capacitors) and, second, it makes the electric field occupy a larger volume of space. If we made a capacitor with two plates a metre apart, it would give a tiny value of capacitance unless the plates were very large indeed.

If one were to make such a cumbersome capacitor and use it as the C in an LC oscillator (such as the Hartley oscillator illustrated in Chapter 15, Figure 15.6), and then carefully measure the energy put in (by the battery) and the amount of heat generated, it would yield an unexpected result. Not all the energy put into the circuit could be accounted for by waste heat. Some of the energy will have escaped – leaked away, if you like – from the area between the plates. This escaped energy is electromagnetic or 'radio energy', and has been **broadcast** away from the circuit. It could be detected by a suitable receiver at some distance from the oscillator.

In order to make a good radio transmitter, we need a circuit that will give the maximum possible 'leakage' of energy from the *LC* oscillator. As you might expect by thinking about the imaginary experiment outlined above, what is needed is a capacitor with the biggest possible gap between the plates, to provide the greatest possible 'leakage' of radio energy. The capacitor would also have to have huge plates to give it a useful capacitance.

We can, in fact, use the largest available 'plate' for one part of the capacitor: we can use the surface of the earth! The other 'plate' has to be a more manageable size, and it is convenient to use a single, long wire; this has the advantage that it can be an excellent 'leaker' of radio energy.

Using the earth and wire as a capacitor is illustrated in Figure 16.1. Is this picture familiar? It shows a radio aerial! There is an optimum length for the aerial wire, if the broadcast energy is to be maximised. Conveniently, it is when the aerial is half, or one-quarter, of the **wavelength** of the oscillator.

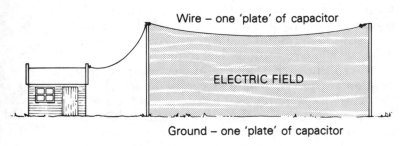

Figure 16.1 A radio aerial.

The term 'wavelength' refers to the physical distance between complete cycles of the broadcast radio wave. Radio waves travel at the speed of light, which is about 300 000 km per second. If the radio waves were produced at the (highly unlikely) rate of one per second, the wavelength would be 300 000 km, still a little on the long side if a convenient aerial is to be one quarter of the wavelength! But radio is broadcast on much higher frequencies than this. Given the frequency of oscillation (f) the formula for working out the wavelength (λ) in metres is

$$\lambda = \frac{v}{f}$$

where v is 300 000 000 (speed of light in m/s). The Greek letter λ ('lambda') is always used to represent wavelength.

As an example, a radio operating at a frequency of 20 MHz ('short wave') would produce a signal of wavelength

$$\lambda = \frac{300\ 000\ 000}{20\ 000\ 000}\ m$$
$$\lambda = 15\ m$$

A convenient quarter-wavelength aerial would be 3.75 m: not too bad, but unsuitable for portable equipment. Fortunately we can trade convenience for efficiency, and use a $\frac{1}{8}$ wavelength or a $\frac{1}{16}$ wavelength aerial.

The length of the aerial will affect the capacitance and this will affect the frequency of oscillation. For a transmitter circuit to work properly, the *LC* circuit must be 'tuned' to the transmission frequency, and the aerial must be the correct length. Radio frequencies between 30 kHz and 20 GHz or so are used. Figure 16.2 indicates the different frequency 'bands', the names given to them, and what they are used for.

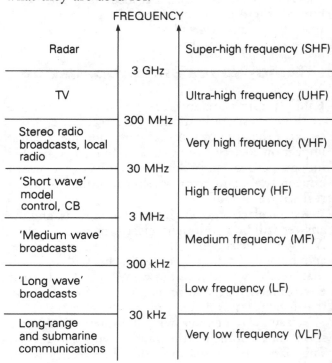

Figure 16.2 The radio spectrum.

Modulation and demodulation

No matter how powerful a signal a transmitter might send out, the signal itself carries no information, other than the fact that it is there or not there. If the radio waves are to carry useful information, such as speech or control signals, the system needs additions. In the early days of radio broadcasting, a simple on–off code was invented, called the **Morse code**. Short and long bursts of radio ('dots' and 'dashes') represent letters of the alphabet, enabling words to be sent (rather slowly) by radio.

There are two common methods of adding information to the radio signal, which is called the **carrier**. Each involves changing the carrier slightly, and both systems are in common use.

Amplitude modulation

The first, and most obvious, is **amplitude modulation** (AM). Amplitude modulation involves nothing more complicated than changing the power, or amplitude, of the carrier in sympathy with the modulating signal. This is easily illustrated in a graph, like the one in Figure 16.3, which shows the carrier being **modulated** with an audio-frequency sine wave. AM has the advantage that it is easy to recover, or **detect** in a radio receiver.

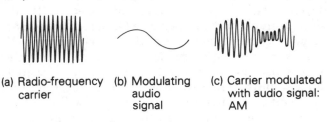

(a) Radio-frequency carrier   (b) Modulating audio signal   (c) Carrier modulated with audio signal: AM

Figure 16.3 Amplitude modulation.

Assuming that the signal received by the receiver is roughly the same as that shown in Figure 16.3(c), we cannot simply feed the output into a speaker. The output is, at audio frequencies, **symmetrical** so that any increase in positive signal is exactly balanced by a similar increase in negative signal and the result is zero: silence. The audio signal can be extracted simply by rectifying the receiver's output, to make it **asymmetrical**. The carrier is then removed with a small capacitor. No explanation other than the graph should be needed!

The circuit is shown in Figure 16.4, along with the waveforms associated with it. Comparison with Figure 16.3 shows how the modulating waveform is recovered more or less unchanged.

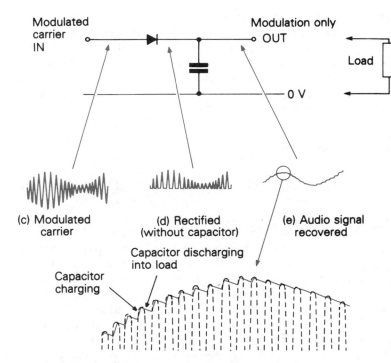

Figure 16.4 A detector circuit designed to recover an audio waveform from an AM radio signal.

Frequency modulation

The second method is known as **frequency modulation** (FM). Instead of changing the amplitude of the carrier in sympathy with the modulating waveform, the frequency is shifted a little higher or a little lower. The amount of frequency shift is very small. Frequency modulation of a carrier is illustrated in Figure 16.5; compare it with the same diagram for amplitude modulation (Figure 16.3).

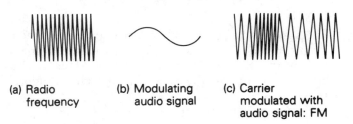

(a) Radio frequency   (b) Modulating audio signal   (c) Carrier modulated with audio signal: FM

Figure 16.5 Frequency modulation.

Detection of an FM signal is much more complicated than detection of AM signals. Various circuits have been developed, and up to the beginning of the 'integrated-circuit era' a circuit known as a **ratio detector** was most commonly used to recover the modulating signal. The functioning of this circuit is quite complicated, and it is now little used. Nowadays a circuit called a **phase-locked loop** is often used. Once again, this is quite complicated and I shall not detail its operation in this book. However, it is nice to know that you can buy a complete phase-lock loop system built onto a single integrated circuit.

Bandwidth

It should be clear that an FM signal broadcasts not on one frequency, but across a narrow range. After all, it is by changing frequency in sympathy with the modulating signal that information is transmitted. It is less obvious that an AM signal occupies more frequencies than that of the carrier, but that is also the case. The 'width' of the range of frequencies used by a radio transmission is referred to as the **bandwidth**. The amount of information that can be transmitted is proportional to the bandwidth. The wider the bandwidth, the more information (and the better the sound quality of a broadcast) that can be transmitted.

In the UK, most of Europe and America – as well as urban areas of most other countries – 'local' radio stations broadcast music on FM at frequencies between 80 MHz and 110 MHz. The high frequency enables a wide bandwidth to be used, and FM is less prone to interference than AM. However, signals at this frequency do not travel very far – a few tens of kilometres – so many stations are needed.

AM radio receivers

The simplest radio receiver consists of just a tuned circuit, a diode, a capacitor, a pair of headphones, and as much aerial wire as possible! This is the 'crystal set' of the 1920s, and is shown in Figure 16.6.

The electric field from the transmitter induces a tiny current in the aerial wire – which needs to be a few tens of metres long – and the LC resonant circuit selects the required frequency. It does this because the LC circuit has low impedance at frequencies other than the resonant frequency, and this 'shorts out' unwanted radio transmissions. At the resonant frequency the LC circuit has a high impedance, so the transmission at the selected frequency appears across the circuit as a small radio-

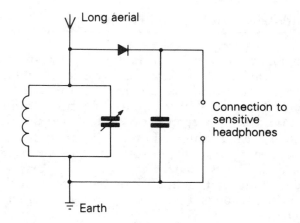

Figure 16.6 *A 'crystal set'.

frequency alternating voltage. This is rectified by the diode to recover the audio signal, using a diode with a low forward voltage drop (germanium is suitable). The capacitor removes the radio-frequency component and a pair of sensitive high-impedance headphones (not so easy to get these days) produce a barely audible output.

The output can, of course, be amplified, but if you were to build a receiver like this and test it, you would discover that the results are unsatisfactory for two reasons. First, the long wire aerial is inconvenient, to say the least. Second, the tuning cannot separate one radio station from the next. Often, several stations will be received together. If we draw a graph of the impedance (apparent resistance to an alternating current) of the LC circuit plotted against increasing frequency, we can see what the problem is. A very powerful station broadcasting on a slightly higher (or lower) frequency than the resonant frequency of the LC circuit will actually give a stronger signal than the one we want.

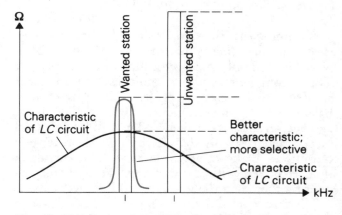

Figure 16.7 A graphical illustration of the way that a receiver needs to be more selective if unwanted powerful radio broadcasts are to be 'tuned out'.

There is a need to make the receiver more **select-ive**, so that it will receive only the station that you have tuned in (Figure 16.7).

Tuned radio-frequency receivers

It is possible to improve selectivity by putting an amplifier *before* the diode demodulator, so that the radio-frequency signal is amplified. If the collector load of the **amplifier** is an *LC* circuit, the selectivity is enormously improved. Two of what are called **tuned RF stages** (RF stands for 'radio frequency') are even better. Figure 16.8 shows a block diagram of such a receiver.

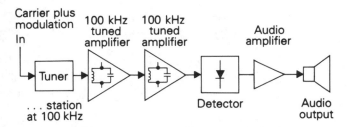

Figure 16.8 A block diagram of an AM receiver with two tuned RF stages.

Think about tuning the radio in Figure 16.8 to different stations, and you will spot the big problem with this sort of design. The problem is that with several stages, each having its own tuned circuit, we have to contrive a means of tuning *every* *LC* circuit simultaneously when changing frequency (station). Such a design would be very cumbersome. Some were made in the 1930s, and were a maze of variable capacitors, inductors, strings and pulleys.

Superheterodyne receivers

There is a much better solution (although quite a complicated one), and that is to use a design called a **superheterodyne** (superhet) receiver. A block diagram is given in Figure 16.9. The incoming signal from the radio-frequency amplifier (the RF amplifier may be left out in very cheap receivers) is **mixed** with a signal from an *LC* Hartley oscillator, known as the **local oscillator**. The local oscillator is tuned to a slightly lower frequency than the selected radio frequency.

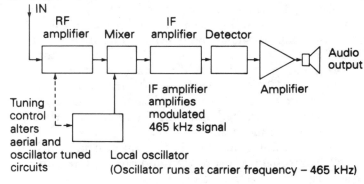

Figure 16.9 A block diagram for a superhet radio receiver.

> Mixing the signal produces a signal that is the difference between the radio frequency and the frequency of the local oscillator, and its amplitude is still proportional to the audio signal that was used to modulate the RF carrier.

The oscillator in most AM receivers is designed to run precisely 465 kHz below that of the received carrier. Here's the clever part: a ganged (double) variable capacitor is used to alter the tuning of both the aerial circuit and the oscillator simultaneously, so that they are always exactly 465 kHz apart. The output of the mixer is therefore always a 'carrier' at 465 kHz, and this carrier is still modulated with the original audio signal.

Now we can use two or three tuned amplifier stages if necessary, without worrying about tuning them to the incoming RF signal. Not only that, but the lower frequency makes design much easier, and the tuning of the stages can be arranged to give a selectivity response close to the ideal. Once the signal has been amplified sufficiently, a detector diode (or, for FM, a suitable discriminator circuit) can follow, with audio amplification last in the line.

The frequency resulting from the carrier and local oscillator inputs to the mixer is called the **intermediate frequency** (IF), and the stages that follow are called **intermediate frequency amplifiers**. They invariably have tuned collector loads: tuned, of course, to the intermediate frequency. The intermediate frequency need not be 465 kHz. Manufacturers produce IF tuned transformers, known as **intermediate frequency transformers** (IFTs), that are preset to the intermediate frequency.

IF amplifier stages invariably use transformer coupling, for the tuned collector load simply needs a secondary winding to make it into a coupling transformer as well, which saves components and makes for an efficient circuit. A typical transformer-coupled IF amplifier stage is shown in Figure 16.10. Note that the IF transformer is shown surrounded by a dashed box. This indicates that it is constructed as a single component; IFTs are mass-produced and are very cheap. The IFT is usually mounted inside an aluminium can, which acts as mechanical protection and also as screening against stray electric fields.

18 frames of film (that is, 18 pictures) every second. Sound movies show 24 frames per second.

The movement on the television screen is created in exactly the same way, and the television transmission is sent out as a series of 'still' pictures at the rate of 25 per second in the UK and many African and European countries, and at the rate of 30 per second in the USA and elsewhere. (The rate is actually half the frequency of the mains supply in the country in question. In the UK the mains supply is at 50 Hz, and in the USA it is 60 Hz.)

Central to the television receiver is the TV tube, illustrated in Figure 16.11. Study the drawing carefully, and compare it with the illustration of the CRT used in the oscilloscope (Chapter 6, Figure 6.5).

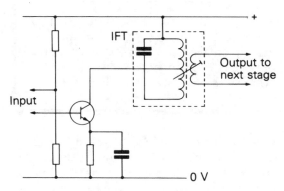

Figure 16.10 A typical IF stage from a small radio receiver.

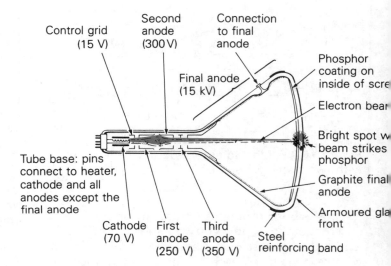

Figure 16.11 A monochrome television tube.

Monochrome television receivers

Having looked at the principles of radio transmission and reception, it is now possible to consider the principles and practice of television. Even monochrome (black-and-white) televisions are extremely complicated in the details of their circuits, and television is a complete subject in its own right. In this book the systems involved will be explained, but without circuit details.

Most people now know that motion picture films produce the illusion of movement by presenting a rapid sequence of still pictures, each one slightly different from the one before it. The eye 'joins up' the pictures and interprets them as a single image in smooth motion: something called 'persistence of vision'. Home movies (silent) show

Apart from the fact that a TV tube is much more widely flared than an oscilloscope tube (to make the TV set shallower), the main difference is that magnetic deflection is used to direct the electron beam, rather than electrostatic deflection. Electrons are easily influenced by magnetic fields, so a suitably designed system of coils, slipped over the outside of the tube neck, can be used to bend the electron beam, causing the dot of light to move up and down and from side to side. Two separate sets of coils are used, for vertical and for horizontal deflection. Designing the deflection coils and the circuits driving them is more difficult for wide-angle tubes, but the convenience (and better sales-appeal) of shallow television receivers makes it worth the effort.

The bright spot is made to scan the front of the tube continuously, in lines running across the screen, moving down from top to bottom in a pattern called the **raster** (see Figure 16.12). The brightness of the raster is reduced during the right-to-left 'return' stroke, the **flyback**. In the UK the screen has 625 horizontal scan lines in the raster; it is slightly different in some other countries. Unfortunately, a scan of this sort would produce visible flicker, even at 25 frames per second, whereas a motion picture film running at the same speed would not. The reason lies in the way the television picture is drawn from the top downwards, whereas the motion picture projector uncovers the whole of each frame virtually at once.

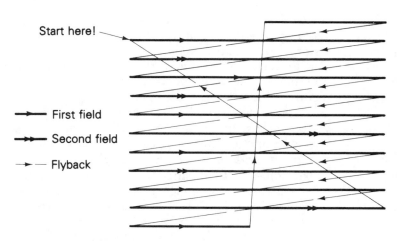

Figure 16.13 Interlaced scanning.

Figure 16.13 Interlaced scanning.

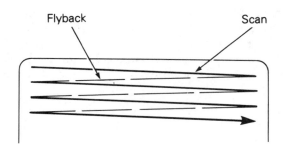

Figure 16.12 A monochrome television screen showing the raster.

To reduce flicker (which has an odd appearance of running from the top down the screen) **interlaced scanning** is used, and the picture is produced in two halves, each taking 1/100th of a second (UK). The electron beam scans the screen as above, but only with 312½ lines. When it gets to the middle of the bottom of the screen (the 312½th line!) it returns to the middle of the top and scans the picture again, but **between** the first set of lines, 'interlacing' the second scan with the first. Using this technique makes the flicker hardly noticeable. Interlaced scanning is illustrated in Figure 16.13, though with an 11-line screen instead of 625, to make the diagram clearer!

It is important to realise that in a monochrome television tube the phosphor on the screen is completely smooth; the lines are merely a product of the scanning that produces the raster.

Two time periods are critical to television: the time that the beam takes to scan the width of the picture (the **line frequency**) and the time taken for one frame (the **frame frequency**). Two oscillators within each receiver generate these frequencies, and they are called, respectively, the **line timebase** and the **field timebase**. The line timebase has to scan the screen $25 \times 625 = 15\,625$ times

every second, so has an operating frequency of 15.625 kHz, and the field timebase scans the picture 50 times every second (twice for each frame; remember the interlaced scanning), and so operates at 50 Hz (the mains frequency – you will see why this is a good idea later).

As the electron beam sweeps across the screen it is modulated (changed in brightness) so that the brightness of the line is changed continuously to make a picture. Figure 16.14 shows how a picture is built up.

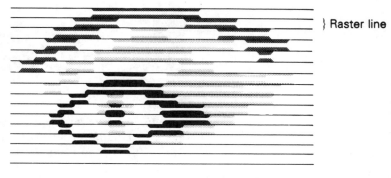

Figure 16.14 The brightness of the spot forming the raster is modulated to provide pictures on the television screen.

Clearly the timebases have got to be exactly in synchronism with the timebases of the cameras at the broadcast studio, or the TV picture will be chaotic, with no recognisable images at all. The transmitted signal therefore includes the information about the timing of the line and field scanning, and circuits in the receiver extract this information and use it to synchronise the two timebases.

Every single sweep of the line timebase is triggered by a pulse in the signal received by the television. A radio-frequency carrier from the transmitter is modulated with a waveform corresponding to the brightness of the line of the TV picture. Figure 16.15 shows part of this waveform, corresponding to two lines of the picture. Compare the waveform in Figure 16.16, which shows two lines of a picture consisting of four vertical stripes, progressively darker from left to right.

The **line-synchronising pulses** each trigger the line oscillator into a single sweep across the picture, or (more commonly) are used to synchronise an oscillator to run at a precise speed. At the end of each field there is a special series of **line-synchronising pulses** that also trigger the **field synchronisation**.

Following the last line of the picture in any particular field there are five equalising pulses, followed by five field-synchronising pulses, followed by another five equalising pulses. The exact function of the equalising pulses is subtle: suffice to say here that it makes the circuits in the receiver simpler. The five field-synchronising pulses are detected by the receiver circuits – they are five times longer than 'ordinary' sync pulses – and trigger the next sweep of the field timebase. Once again, a synchronised oscillator is used. This prevents the picture from collapsing entirely if the signal is lost momentarily.

Following the pulse sequence for field synchronising there are a further 12½ 'blank' lines that are not used for the picture. These are put in to give the electron beam in the tube time to fly back to the beginning of the next field at the top of the picture. When all these various pulses and blank lines are taken into account, a total of 20 lines are 'lost' for each field: that is, 40 for each frame. For a television system like that used in the UK, having 625 lines per frame, only 585 lines actually appear on the screen.

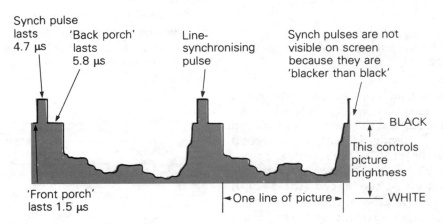

Figure 16.15 The waveform of a television picture broadcast. This is the waveform that is recovered from the AM signal received by the television.

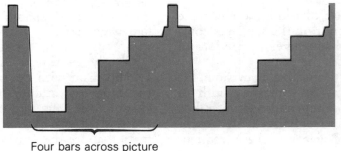

Four bars across picture

Figure 16.16 Two lines of a television picture showing four vertical bars of increasing density.

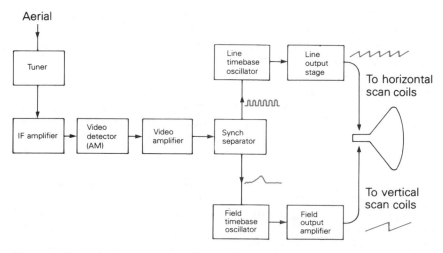

Figure 16.17 A system diagram of the synchronisation and scanning circuits of a monochrome television receiver.

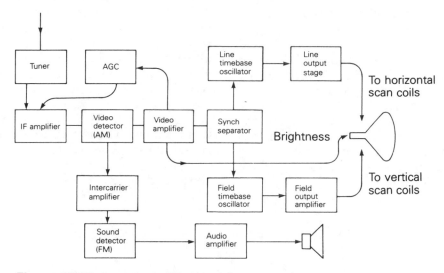

Figure 16.18 A system diagram of a monochrome television receiver.

A system diagram for the line and field scan of a television receiver is given in Figure 16.17.

Television sound

The signal for TV sound is transmitted on a completely separate carrier; in the UK it is 6 MHz higher in frequency than the carrier used for the vision.

A complete television receiver

A system diagram for a complete TV receiver is shown in Figure 16.18.

Colour television receivers

The problems facing the designers of the colour television system were formidable. The television system was well established with monochrome receivers, and it was important that the colour television system used would be cross-compatible.

That is, a black-and-white receiver should be capable of receiving a colour transmission and would reproduce it correctly (in black and white); and at the same time, a colour receiver should reproduce a black-and-white transmission as well as a monochrome receiver. The extra information required for the colour receiver therefore had to be 'fitted in' round the existing signal, and in such a way that it would not interfere with the operation of a monochrome receiver. The colour signal also had to be squeezed into the available bandwidth, which had been established as 8 MHz.

Fortunately, the eye is far less sensitive to colour than it is to brightness, and it has proved possible to transmit the colour information over a rather restricted bandwidth: the colour on a colour television is actually rather fuzzy. But if the brightness is controlled with a relatively wide-bandwidth signal – giving a sharp picture in terms of brightness – the resulting picture is perfectly acceptable. Figure 16.19 shows a waveform corresponding to one line of the same four vertical stripes illustrated in Figure 16.16 above, but with colour information added. The colour signal is actually coded during the vision information period, and at first sight seems to be likely to interfere with a monochrome receiver. But remember that the frequency of the colour signal – the **chrominance** waveform – is high, at 4.43 MHz, and would not be resolved by the circuits of the monochrome receiver, which would average the signal.

A complicated method known as **quadrature modulation** is used to retrieve information about three colours from the frequency-modulated chrominance subcarrier. (The waveform is called a 'subcarrier' because it is a carrier waveform, derived from modulation of another carrier of higher frequency.) The complete range of colours can be reproduced from just these three colour signals, as we shall see.

Slight changes in the transmitted waveform caused by weather conditions, reflections from aircraft, etc., can affect the colour information, causing a drift of the whole colour spectrum of the picture towards red or blue: not a desperate problem, but enough to notice. Despite this problem, this system was adopted in the USA, the first country to have a broadcast colour TV network. The system is called the NTSC system, after the US **National Television Standards Committee**.

The system adopted by almost everybody else, with the benefit of learning from the pioneers' mistakes, is known as PAL, which stands for **phase alternation by line**. This overcomes the colour shift of NTSC by the ingenious method of turning round the modulation system on alternate lines. Colour shifts still occur, but alternate lines are changed towards blue and then red, and blue and then red, etc. The eye merges the colours and the overall impression is correct.

Even with this improvement, very large colour shifts in the picture still looked odd, a little like a

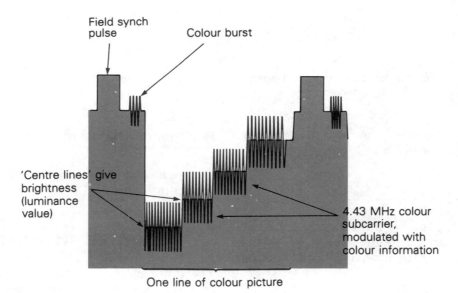

Figure 16.19 The waveform of one line of a colour television picture. The waveform is for a picture consisting of four vertical bars, like the one in Figure 16.16.

multicoloured venetian blind, so a modification of the PAL system, known as PAL-D was introduced. PAL-D actually adds the colour signals of alternate lines electronically, by storing a whole line in a device called a **delay line** while waiting for the next line. The colour on the screen is always the product of two alternately coded signals, and the result is effectively perfect colour stability. The French use a rather different but equally effective method called SECAM, standing for **séquential couleur à mémoire**.

Combining colours

The information recovered from the transmitted signal is about three colours only, but it is possible to use the three colours to make every colour in the visible range, in just the same way that an artist can make every colour (if he wants to) by mixing primary colours. The three primary colours of the colour TV are mixed in different proportions to produce the whole range of every colour that we can see. The three colours used are red, green and blue. Mixed together in equal proportions, these make white (white light is 'all the colours of the rainbow' mixed together), and other colours can be made by mixing two or three in different proportions.

For example, green and red mixed together produce a bright yellow. This process is called **additive mixing**, and is not quite the same as the mixing of artist's colours, which is called subtractive mixing. The artist uses pigments that reflect only certain colours from the white light falling on them: they

subtract colours. In the TV tube, colours are mixed: they are added together.

There are three factors that control the picture on the screen: **brightness**, **hue** and **saturation**. What do these mean? 'Brightness' is straightforward, and on the screen is controlled by the increase or decrease of all three colours simultaneously. 'Hue' (colour) is controlled by the balance of the three colours, as described above. 'Saturation' refers to the 'strength' of the colour: the amount of white light that is added to the basic colour. Red, for example, is a saturated colour. Pink is red mixed with white, and could be said to be a less saturated red.

White light is not of course produced by the TV receiver, so saturation is in fact controlled by the difference between the brightness of the three primary colours; it is convenient to think of it as the colour mixed with white, however.

Colour picture tubes

The colour TV tube is a development of the monochrome tube illustrated in Figure 16.11. The overall shape is the same, but there are three separate electron guns. In the most popular type of tube, the **slot-mask tube**, the three electron guns are arranged in a row, horizontally. The guns produce three electron beams, one for each colour, and although the beams are all deflected together by the vertical and horizontal scan coils, the brightness of the beams can be controlled separately.

All three beams therefore scan the front of the tube together, to produce a raster. Obviously, it is

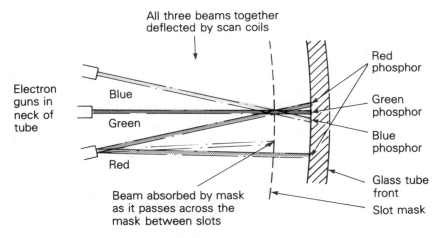

Figure 16.20 The principle of the slot-mask colour television tube (not to scale).

not possible to have coloured electron beams, so the colour is produced by **phosphors** on the front of the screen. Phosphors can be made almost any colour, and the right shades of red, blue and green can be produced easily, if expensively. The raw materials for the coloured phosphors actually contribute quite a lot to the price of a colour TV tube.

The trick is getting the 'red' beam to affect only the red phosphor, the 'blue' beam to affect only the blue phosphor, and the 'green' beam to affect only the green phosphor. This is done by simple geometry. Figure 16.20 shows the system diagrammatically, viewed from the top of the tube.

The slot-mask is fixed firmly in place behind the tube, and the relationship of the positions of the slots and of the phosphor strips on the back of the tube is such that the 'red' electron beam can fall only on the red phosphor, etc.

Figure 16.21 is a perspective sketch of the tube, from which you can see that the slots are quite short, and are 'staggered' to produce an interlocking pattern. This arrangement is used because it is physically strong – great rigidity of the mask is clearly important – and because it gives a reasonably large ratio of 'slots' to 'mask'. The more transparent the mask is in this respect, the better, for electrons that hit the mask rather than go through the slots are just wasted power, and serve only to heat the mask. The larger the slots can be made in

relation to the mask, the brighter the picture will be.

The slot-mask principle is used in the popular **precision in-line** (PIL) tube, which is manufactured complete with scan coils as a single unit. This, together with extremely sophisticated scan-coil design (produced in the first place with computer-aided design techniques) has led to a tube that is very simple to use, all the most critical alignments having been built in at the manufacturing stage.

There is no space here to give more than the briefest glance at colour television receivers. The circuitry is, in both theory and in practice, very complicated, and extensive use is made of special-purpose integrated circuits to reduce the 'component count' and to make the receivers easy and cheap to manufacture. In real terms a colour television costs far less than its monochrome counterpart of 30 years ago, and gives incomparably better results.

> ▲ Television servicing, especially colour television servicing, is highly specialised, and should never be attempted by the inexperienced or untrained. The voltage on the final anode of a colour receiver will be more than 25 kV and is usually lethal.

Before leaving the subject of colour TV, try to get hold of a pocket magnifying glass. Take a close look at the surface of a colour TV tube, first with the receiver turned off, then with it turned on. You can see the coloured phosphors very clearly.

Television cameras

Clearly, the image that is re-created by the television receiver has to be photographed in the first place. Modern television cameras are small and reliable, thanks to the widespread use of ICs. Much the same technology is used in 'camcorders' (combined cameras and video tape recorders).

A simplified drawing of a television camera is shown in Figure 16.22. The zoom lens assembly projects an image of the subject onto a device known as a **charge-coupled detector** (CCD for short). The CCD is a development of integrated

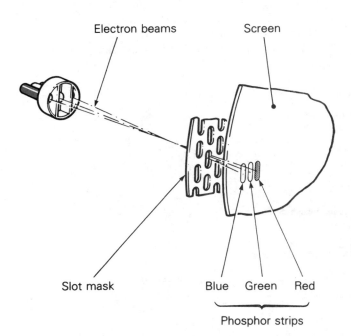

Figure 16.21 A perspective drawing of a slot-mask colour TV tube (not to scale).

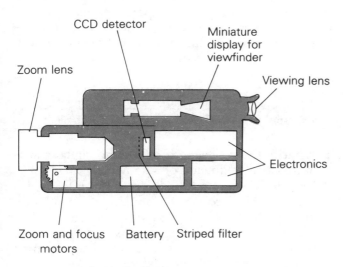

Figure 16.22 A television camera.

circuit technology, and consists of an array of many thousands of individual light-sensitive FETs, arranged in horizontal lines. CCDs have completely superseded older devices based on thermionic technology (they had names like 'image-orthicon' and 'plumbicon'), but unfortunately details of the way CCDs work is well beyond the scope of this book.

A **striped colour filter** is positioned in front of the CCD. It separates the three colours. Broadcast cameras usually have three-colour filters, but smaller cameras have two-colour filters (green and magenta) and a system to calculate the missing third colour (cyan). This enables a smaller and cheaper CCD to be used.

The CCD is 'scanned' electronically at exactly the right speed, so that the 585 lines of the picture can be transmitted and received by television receivers. Electronic systems add the required sync pulses.

Figure 16.23 shows, in general terms, how the signal gets from the camera to the receiver, in this case via a satellite link.

◼ CHECK YOUR UNDERSTANDING

● Radio behaves like waves, travelling at the speed of light.
● Although we talk about 'radio waves' radio

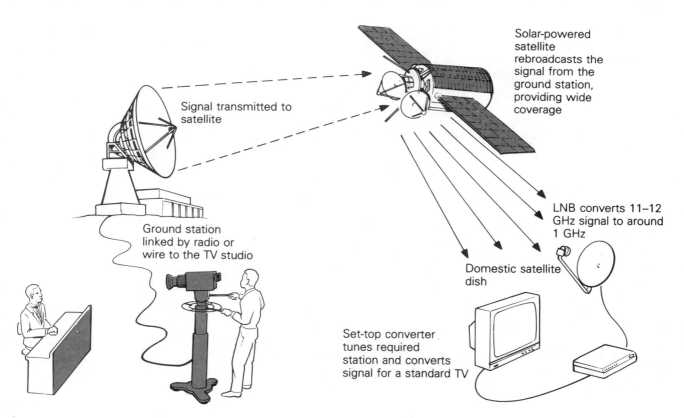

Figure 16.23 From the TV studio camera to your television.

cannot really be waves as it will travel through a vacuum. It is just useful to think of it that way.

● A radio signal is measured in either **frequency** or **wavelength**. The two are just different ways of expressing the same thing.

● To carry information, a radio signal must be **modulated** with an audio (or other) signal.

● A radio signal can be modulated in two different ways. It can be changed in power (**amplitude modulation**, AM) or in frequency (**frequency modulation**, FM).

● The signal must be **demodulated** by the receiver.

● Most modern radio receivers use the **superheterodyne** principle.

● Television receivers (monochrome) work by **scanning** the picture tube in a number of horizontal lines to build up a **raster**.

● Television raster scanning is done very quickly (50 Hz or 60 Hz) by a single spot of light. The brightness of the spot varies to build up a monochrome picture.

● We see a complete picture because **persistence of vision** means that our eyes cannot respond quickly enough to detect very rapid scanning by the spot of light.

● Colour television receivers work in much the same way, but have three separate spots that produce red, green and blue on the screen. By controlling the relative brightness of the three beams any colour can be produced.

REVISION EXERCISES AND QUESTIONS

1 What is the wavelength of a radio signal transmitted at a frequency of 12.5 MHz?
2 Explain what is meant by i) amplitude modulation and ii) frequency modulation. iii) Why do you think the AM system was invented first?
3 The superheterodyne system is widely used in radio receivers. What is its great advantage over earlier systems like the TRF receiver?
4 Why is interlaced scanning used in television?
5 Colour television picture tubes always have three electron beams. Why three?
6 Most superheterodyne radio receivers have an intermediate frequency of 465 kHz. Why do you think this is, when (for example) 400 kHz or 473 kHz would clearly work just as well?

Construction methods, fault-finding and repairs

Introduction

In this chapter you will learn about **soldering**, which is the basic technique used for the assembly of all electronic systems. Soldering is more of a craft than a science, and you should practise as much as possible. You will find that you will get better and better. Soldering seems difficult and fiddly at first (and the first soldered joints you make will quite likely fall apart!) but you will soon realise that it is a very quick and effective method of making electrical connections and supporting small components.

The concept of **maintenance** is also introduced in this chapter, along with the basic principles of repairing electrical and electronic equipment.

Assembly tools and techniques

Soldering

Components are connected together in circuits by soldered joints. Solder is an alloy of lead and tin, the proportions of which vary according to the application. For electronics work a mixture of 60 per cent tin and 40 per cent lead is usual.

If solder is heated to its melting point it will, if applied to a variety of different metals, amalgamate with the metal surface and produce a joint with high electrical conductivity and a good mechanical strength. The strength is limited by the relatively poor tensile strength of the solder itself (neither lead nor tin are very strong metals). Solder has a low melting point, lower than that of either tin or lead. The melting point varies according to the

exact composition of the solder, but ordinary 60/40 tin/lead solder melts at 188 °C.

The temperature required to melt the solder is also sufficient to oxidise many metals and form a layer of oxide that will prevent the solder from 'sticking' to the surface. Solder intended for electrical work is therefore made in the form of a hollow wire, with one or more cores of **resin flux**, a chemical mixture that will dissolve the oxide film at soldering temperatures. A cross-section through two different makes of electrical flux-cored solder is shown in Figure 17.1. Solder can be obtained in bar form, without the flux core.

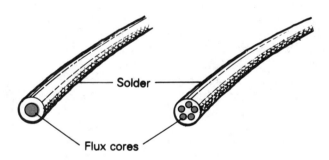

Figure 17.1 Solder.

For prototype and repair work, soldering is carried out using a **soldering iron** or **soldering gun**. The **bit** or tip of the iron is heated electrically to between 350 and 420 °C. To make the joint, the bit of the iron is applied to the two surfaces to be joined, and the wire solder is applied to the heated joint. The solder melts, allowing the flux to run over all surfaces and clean them. The solder also helps to conduct heat from the iron onto the surfaces, and when the temperature is high enough it amalgamates with them. The iron can now be removed and the finished joint allowed to cool.

The means of heat generation is different in soldering and soldering guns. A soldering gun is illustrated in Figure 17.2. The gun is basically a transformer. The secondary winding has a few turns of very heavy copper wire or bar, and generates a very heavy current at low voltage. The bit of the gun is made of copper and, being substantially thinner than the transformer secondary windings, has a relatively high resistance. The energy generated by the transformer secondary winding heats the copper bit rapidly, until the soldering temperature is reached.

excellent insulating properties and low capacitance, and provides a good measure of safety when soldering sensitive circuits. A sectional drawing of a ceramic-shafted soldering iron is shown in Figure 17.3.

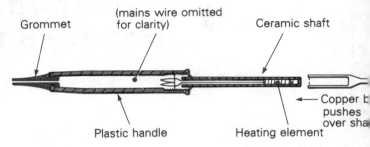

Figure 17.3 A ceramic-shafted soldering iron, used for light soldering jobs; essential where circuits that are sensitive to electrostatic voltages are being soldered.

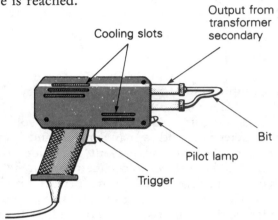

Figure 17.2 A soldering gun: ideal for heavy soldering and some kinds of repair work.

Soldering guns have the advantage that they are quick to heat up – only a few seconds – and cool down rapidly after switching off. The gun is on only when the trigger is held down. Soldering guns are also powerful, and are useful for soldering larger components or small metal sheets. A typical soldering gun produces heat output of some 50–150 watts. The main disadvantage is that heat regulation is not particularly good, and it needs a certain amount of skill to keep the bit at the proper temperature. Also, the bit is large and rather clumsy, making the gun unsuitable for fine work.

Soldering irons are used for most work. A soldering iron uses a heating element to bring the bit up to the required temperatures. A small iron for electrical work would have an output of 7–25 watts, depending on size; for delicate work (such as integrated circuits) about 10 watts is sufficient.

Some types of integrated circuit are damaged by the slightest electrical leakage of high voltage from the mains supply, through the insulation of the iron (or through internal capacitance). To reduce the danger, **ceramic-shafted soldering irons** are available, in which a thin ceramic sleeve insulates the bit from the rest of the iron. The sleeve has

Soldering is something of an art, and needs to be practised. Typical bad joints are caused by insufficient heat, dirty surfaces, insufficient solder or persistent reheating of a joint in an attempt to get it to stick. A selection of problems is shown in Figure 17.4.

Some metals – aluminium, for example – cannot be soldered by normal techniques. Metals that can be soldered easily include gold, silver, tin, copper and lead. Gold is often used as a very thin plating over a copper or iron component or connecting wire, to improve solderability.

Desoldering

To remove a soldered component is usually easy, but not always. Usually a component can be heated up with the soldering iron and the connections pulled apart. In some cases this is not practicable: for example, where an integrated circuit is soldered directly into a printed circuit. All the pins – sometimes as many as 40 – would have to be heated up simultaneously before the component could be removed from the board! To assist in removing this sort of component, a **solder sucker** is used. A solder sucker is illustrated in Figure 17.5.

The principle is very simple, rather like a back-to-front bicycle pump. The plunger is pushed down against the spring, where it is locked by the trigger. The soldered joint to be released is then heated with a soldering iron, the nozzle of the sucker applied to it, and the trigger pressed. The spring forces the plunger up the tube, sucking the

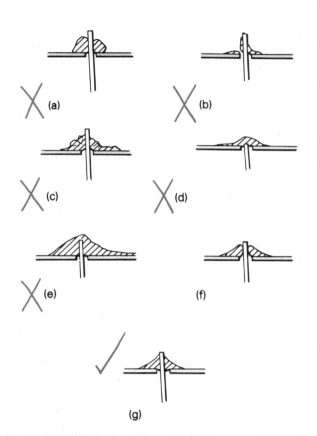

Figure 17.4 Some soldered joints in cross section.
(a) Soldered joint too cold. (b) Insufficient solder.
(c) Solder reheated too often. (d) Wire not far
enough into hole. This results in a joint that is
mechanically weak. (e) Too much solder forms a
'bridge' to the next part of the circuit board.
(f) A 'dry joint'. The solder has stuck to the board,
but not to the wire, perhaps because the wire was
dirty: a difficult fault to detect, and can develop
a high electrical resistance. (g) Correctly made
joint.

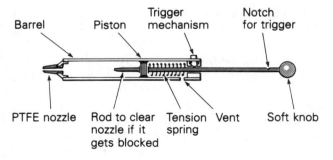

Figure 17.5 A solder sucker. The nozzle is made
of polytetrafluoroethylene (PTFE), a plastic that
is resistant to heat and to which molten solder
will not stick.

molten solder up after it and also cooling it
instantly. Some solder suckers have a double-
sprung plunger to prevent it from flying out and
hitting you in the face when you press the trigger.

Hand tools

A range of small hand tools is used for electronics
work, but there are few specialised tools. A variety
of long- and short-nosed pliers should be available,
as well as small and medium side-cutters. Several
screwdrivers, both flat and cross-point, are essential. The only specialised tools that are often found
are wire strippers, for removing the insulating plastic from connecting wires; trimtools, small plastic
screwdrivers for adjusting the cores of inductors;
and perhaps a miniature electric drill and a magnifying glass.

In some high-quality prototype work, an
assembly technique known as **wire wrapping** is
used; there are a number of special devices that
are used for wrapping and unwrapping wires.
Briefly, wire wrapping consists of making a connection by wrapping a wire tightly round a hard
square post with sharp corners. This results in a
joint with good mechanical and electrical properties, made without heat.

Repairs

It is important to know your own limitations!
Never attempt to repair anything unless you
are pretty certain that you can do the job.
Remember, you are only just beginning your
studies in electronics.

Industrial maintenance and repairs

There is a big difference between **industrial and
commercial** repairs and **domestic** repairs. An
industrial system that fails needs to be repaired
quickly because it could be costing a lot of money
while it is not working (for example, in a factory).
This is not the case if your domestic radio breaks
down.

The object of industrial/commercial repairs and maintenance is to keep the equipment working for as much of the time as possible. Some pieces of equipment (or subsystems) are more reliable and others, and the reliability is measured as the **mean time between failures** (MTBF for short); this is – on average – how long it will operate between each fault.

The length of time that something takes to fix is the **mean time to repair** (MTTR for short). This includes the whole time between the failure and getting it working again, not just the actual time taken to work on it. For example, it includes the time taken for the repair man to get to the site, and the time taken to get any necessary parts.

The **availability** (A) of the equipment to the user is calculated as

$$A = \frac{MTBF}{MTBF + MTTR}$$

This is the time during which equipment is available to be used. As you can see, it is adversely affected by poor reliability, and by taking a long time to repair.

Preventive maintenance

The object of preventive maintenance is to prevent an item of equipment from failing. An everyday example is the motor car. Cars are (usually) serviced regularly to prevent them from breaking down somewhere inconvenient. The object is to replace parts that are wearing out *before* they actually fail.

It is less obvious what to do about electronic equipment. Since modern electronic devices are very reliable, there is an increasing likelihood that an engineer, in checking a system, will actually *cause* a fault. This is particularly true of systems that are complex electronically but simple mechanically. Computers are an example: preventive maintenance is limited to checking mechanical components like disk drives.

In practice, only safety-critical equipment will normally be subject to preventive maintenance.

Corrective maintenance

Corrective maintenance involves repairing things after they have failed, and is the usual way of doing things.

Domestic repairs

Before an engineer begins works on a domestic item, he needs to be aware that most electronic (and other) equipment has a 'design life'. Nobody – not even a radio – lives for ever.

Is it worth repairing?

The first thing to consider is whether a repair is *worth* doing or not. If you are faced with a radio, tape recorder, or other item of equipment that is severely damaged, worn out, or easily replaceable, you have to make a decision whether or not a repair is worth the time and money. Some items – such as cheap transistor radios – are unlikely to repay the cost of repair: they could easily cost more to fix than they are worth.

Some items are so costly and difficult to repair that it may not be worth the effort. For example, an old television that needs a new picture tube is probably not worth mending. The high cost of (and often difficulty in obtaining) a new tube means that it would be better to throw the set away and get a new one.

Can you discover how it works?

It is often necessary to have the manufacturer's manuals before you can attempt a repair. Electronic items such as computers, televisions and radios (other than the simplest) can be very difficult to fix without the maker's service manual. If this is not available, and is likely to be hard to get, you may be well advised not to attempt a repair. A piece of equipment with a lot of integrated circuits – especially if they are purpose-made by the manufacturer – may be very difficult to follow.

Can you repair it?

Some circuit boards and components are *very* small; repairing them is more like watch-making than electronics. Camcorders are often like this. In the following sections of this book I shall consider the most common kinds of circuit assembly, but if you come across **surface-mounted** components, you will almost certainly be unable to work on the board without specialised equipment. In such cases the manufacturers may well provide a 'service exchange', or you may just have to give up and declare the item 'beyond economic repair'.

It is important to make a professional judgement, just like a medical doctor does. If the patient is incurable, it is best to say so.

Fault-finding

A failure will cause a particular set of symptoms, and the engineer's job is to discover which component is at fault. This is not always easy, as the symptoms may be very indirect. One of my own experiences is a good example. The fault was in a motor car: hot water was trickling over the driver's legs, under the dashboard. The cause turned out to be an electronic failure in the alternator! The sequence of events was as follows. A chip in the regulator circuit in the alternator had failed. Instead of producing about 14 V, the alternator output rose to around 25 V at speed. This very high charging current caused the car battery to overheat and, eventually, to boil. Hot acid sprayed out of the battery's safety vent. The acid dissolved part of the interior heater control valve, allowing water to escape and drip through the dashboard.

The same kind of obscure trains of events can occur in electronic systems. A failed resistor can burn out a transistor, which damages an IC, and so on. A logical approach helps track down the root of the problem.

A systems approach

When attempting to locate a fault, you should try to break the problem down into a 'block diagram' to help pin-point the problem. For example, most pieces of equipment have a distinct power supply section (see Chapter 14). It may even be on a separate circuit board. A good starting point might be to check that the output(s) of the power supply are correct.

You should use a **sequential** approach to fault-finding. Move from part to part of a system in a logical fashion, trying to discover which part(s) of a system are in working order. Here is an example of the way in which one might tackle a fault in an audio amplifier.

> Q. What's the symptom?
> A. No sound.
>
> Q. Anything obviously wrong: burnt or overheated components, detached wires?
> A. No.
>
> Q. Is the PSU working?
> A. Yes; voltage seems normal (refer to manual if possible).

> Q. Are voltages in the output stage OK (in a complementary stage the voltage on the output capacitor should be about half the supply voltage)?
> A. Yes.
>
> Q. TEST: apply a signal to the input (a signal generator, oscillator or audio source). Is there a signal:
>   1. At the input of the output pair?
> A. No.
>
> Q. 2. After the volume control?
> A. No.
>
> Q. 3. Before the volume control?
> A. Yes.

The fault is therefore somewhere in the volume control part of the amplifier. Check that the potentiometer is OK (that all three terminals are connecting to the track) then look at – and replace as necessary – the other associated components. Check all soldered joints carefully.

It is vital to adopt a logical and methodical approach to testing and fault-finding, if only because this will minimise the amount of time that you spend on the job! Domestic equipment is not usually as well made as industrial equipment, and 'ordinary people' may not be aware of what to look for when something electronic goes wrong.

Start by looking for obvious problems that the user may have missed. If the equipment is battery-powered, is the battery all right? Use a voltmeter to check it under load; don't take someone's word for the fact that the battery 'is new'. Mechanical components fail more often than electronic ones, so look carefully at switches, battery clips, plugs and sockets, and any mechanical linkages that may affect the operation of the equipment.

An item that has been dropped (although the owner may deny most vehemently that it has!) will often exhibit a whole range of symptoms associated with wires pulled off, soldered joints, and even components fractured.

Printed circuit boards

Almost all electronic circuits are built on **printed circuit boards** (PCBs for short). A PCB consists of a fibre sheet 1 mm or 2 mm thick, on one side of which there is bonded a thin layer of copper. During the manufacturing process most of the copper is etched away to leave 'lands' of copper that form the interconnecting wiring of the circuit.

Small holes drilled through the board allow the connecting wires of resistors, capacitors, transistors, etc. to be passed through from the 'front' (non-copper) side of the board and soldered to the copper on the 'back'. Figure 17.6 shows some components mounted on a PCB.

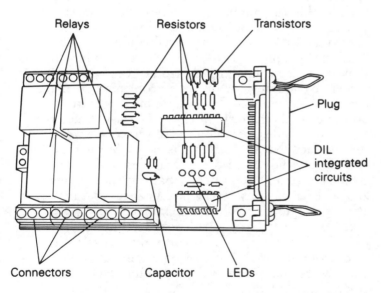

Figure 17.6 Various components mounted on a printed circuit board.

A PCB is a very convenient way of supporting small electronic components and connecting them together at the same time. Circuits built on PCBs are – with a few exceptions – easy to repair.

The first exception is if you come across a **flat pack**. When ICs are mounted on a PCB they are usually fixed in the same way as other components, by passing the connecting pins though holes in the board and soldering them on the back. Sometimes an IC is even fitted into a **DIL socket**, which makes changing it very easy. If it isn't, you have to use a **solder sucker** (see Figure 17.5 above) to remove solder from all the pins in order to extract it from the PCB. Flat packs, however, are soldered directly onto the copper side of the board and can be quite difficult to remove. All the connecting leads have to be heated and lifted at the same time. Flat packs are often very small and fragile, and may be glued to the PCB. If a flat pack fails, it is quite usual to throw away the whole board, or even the whole subassembly (for example, a complete disk drive with its associated electronics). Fortunately, these ICs are extremely reliable!

The second exception is **surface mounting**. When a circuit has to be very small, tiny resistors, capacitors and transistors are soldered directly to the copper of the PCB, without connecting leads. At this stage, don't attempt repairs to any PCB with surface-mounted components.

Be careful how you deal with any PCBs that have ICs fitted; many ICs are sensitive to electrostatic voltages and can be destroyed simply by touching the pins under the wrong conditions.

Signal tracing

The technique described in the question-and-answer panel above used a method called **signal tracing**, in which a signal is followed through the system to discover which part of the equipment is at fault. It might be convenient to use a signal that is already present, such as an audio input from a tape deck or record-player if testing an amplifier. Alternatively, a **signal generator** can be used to provide an audio tone or radio-frequency input.

It is best to trace the output with an oscilloscope. This will not only show the presence (or absence) of the test signal, it will also indicate whether it has been distorted. Because the graticule of an oscilloscope can be used to measure voltage, you can use the instrument to check the gain of an amplifier stage, by comparing the peak voltage of the input waveform with that of the output voltage of the stage.

Signal injection

A slightly different technique can be used: that of feeding a signal into the input of each stage – each 'block' of the system – in turn, and checking the output. In order to do this, you need to know what kind of input (voltage, frequency, impedance) the stage is expecting, and have the ability to simulate it with a signal generator.

Knowing the limits . . .

If you use a methodical and logical approach to testing and fault diagnosis you will be able to repair a good number of electronic items. But you must remember that many modern electronic systems are very complex, and can be diagnosed and repaired only with access to the manufacturer's manuals and circuit diagrams.

Many items (for example, some PSUs) will have various safety circuits designed to protect other parts of the system in the event of a failure; these can be particularly hard to repair, and the makers may recommend replacing the whole subsystem if anything goes wrong with it.

Computers based on the IBM PC are designed this way. All the major systems are on separate boards – printer ports, serial ports, video, extra memory, network connector, modem – and if a fault develops in any of these, the board is replaced and the old one discarded. Repairs would be difficult and more expensive than the replacement board.

I began this section by reminding you that you should know your own limitations and never attempt to fix anything unless you are certain that you can do so. I'm ending this chapter the same way.

■ CHECK YOUR UNDERSTANDING

● **Solder** is an alloy of lead and tin.
● Solder is used to make the connections between electronic components and to fix them to their mountings (usually a **printed circuit board**).
● Because FETs and some integrated circuits are damaged by high voltages, even electrostatic voltages, you should use a **ceramic-shafted** soldering iron when working on boards containing such components.

● A **desoldering** tool should be used to remove components from a circuit board.
● **Never** try to repair anything unless you are confident that you know what you are doing.
● A logical (systems) approach should always be used in fault-finding.
● At this stage, do not attempt to work on any printed board that involves **flat packs** or **surface-mounted components**.

1 How would you set about removing a printed circuit mounted relay from the PCB? This component has five terminals, all soldered through the PCB.
2 The circuit shown below doesn't work.

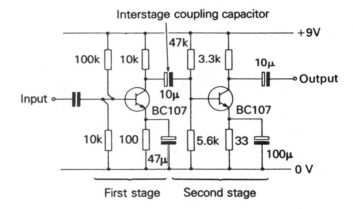

Interstage coupling capacitor

First stage   Second stage

There is no output and the second-stage BC107 is very hot. You replace it and the circuit still doesn't work and the replacement transistor gets very hot, too. Which component(s) should you look at next?
3 Someone brings you a transistor radio to fix. The output is very distorted and the volume is low. The distortion is worse when you turn the sound up. What do you look at first?
4 A (mains-powered) tape recorder is in for repair. The sound is very muffled and the volume is poor. What do you look at first?
5 A tape recorder brought in to be repaired is working quite well but is getting so hot that the plastic cabinet is beginning to melt! You take it to pieces and discover that the heat seems to be

coming from the transformer in the low-voltage power supply circuit, which is a simple unregulated design with a bridge rectifier (which seems fairly cool) and a big smoothing capacitor that also has been getting hot. What fault should you suspect?

6 Someone brings you an old IBM PC style computer to look at. The symptom is that the machine is completely dead. You suspect there is something wrong with the power supply. What should you do next?

Digital logic

Introduction

You may already know that digital electronic computers work by using specialised electronic systems that can count, do calculations, and remember numbers.

An initial study of the way that **digital logic** circuits work could easily fill another two books the size of this one, so all I shall do here is to present an 'overview' in terms of **systems** without going into any details of device physics, 'families' of ICs, or types of circuit.

Your syllabus may not call for a knowledge of digital logic, but even if it does not, I should like you to at least read through this chapter, as it will help give you an insight into **digital systems**, which are becoming more and more important.

Logic systems

We begin by considering a simple machine, such as a coffee and coke vending machine. Before it delivers a cup of coffee, the machine needs two things to happen: (1) you put some money in, and (2) you press the button to select 'coffee'. We could write this down as a logical statement:

money AND *coffee button* = *coffee*

This can be represented as a diagram (Figure 18.1). The rectangle represents a **logic element**.

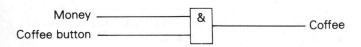

Figure 18.1 An AND logic diagram.

What about the coke? If we want to design a system that gives coke as an alternative to coffee, we can represent this with the diagram in Figure 18.2. Notice that one of the inputs (money) is common to both logic elements:

money AND *coffee button* = *coffee*
money AND *coke button* = *coke*

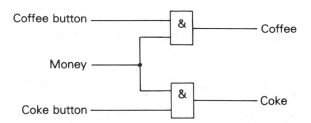

Figure 18.2 A logic system with two AND elements.

If I were to manufacture and sell this machine, I would get a lot of complaints. Can you see why? Sooner or later someone will discover that

money AND *coffee button* AND *coke button*
= *coffee* AND *coke*

We could have predicted it if we had written down all the possible combinations of inputs to the system as a table. Such a table, in which 'yes' is represented by a **1** and 'no' is represented by a **0**, is called a **truth table**. Figure 18.3 shows the truth table for our machine.

The 'two drinks for the price of one' problem is solved by adding two extra elements, as in Figure 18.4. Think of what is happening in this diagram in terms of 2s and 0s. The logic elements are labelled G_1–G_4. Notice that G_3 has a ○ symbol on its output. This symbol means **negate** and reverses the output: if it's a 1 it becomes a 0 and if it's a 0

Inputs			Outputs	
Coffee	Money	Coke	Coffee	Coke
0	0	0	0	0
0	0	1	0	0
0	1	0	0	0
0	1	1	0	1
1	0	0	0	0
1	0	1	0	0
1	1	0	1	0
1	1	1	1	1

Figure 18.3 Truth table for the badly designed drinks vending machine.

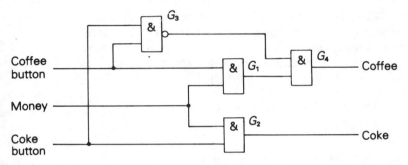

Figure 18.4 A better logic system for the vending machine.

it becomes a 1. Ignore G_3 and G_4 for the moment, and the diagram is quite like the one in Figure 18.2, except that the output of G_1 goes to an input of G_4. All the while the other input of G_4 is 1, its output will only go to 1 when the output of G_1 is also 1, signalling 'coffee'. Take a few moments to trace through the possibilities on the diagram to reassure yourself of this.

Inputs			Outputs	
Coffee	Money	Coke	Coffee	Coke
0	0	0	0	0
0	0	1	0	0
0	1	0	0	0
0	1	1	0	1
1	0	0	0	0
1	0	1	0	0
1	1	0	1	0
1	1	1	0	1

Figure 18.5 A truth table for the better vending machine in Figure 18.4.

Now, when our 'clever' customer presses both buttons, both inputs to G_3 are 1, and the output (don't forget that ○ !) goes to 0, preventing the delivery of coffee. All he gets is coke.

The truth table for this system is shown in Figure 18.5.

Logic gates

The diagrams above are **logic diagrams** and the logic elements are more usually called **logic gates**. There are six basic types in common use, as follows.

AND/NAND gates

Figure 18.6 shows the two kinds of gate that we have already touched on, the AND gate and the NAND (negated-AND) gate, along with their truth tables.

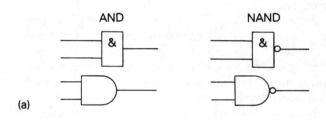

(a)

AND		
Inputs		Output
0	0	0
0	1	0
1	0	0
1	1	1

NAND		
Inputs		Output
0	0	1
0	1	1
1	0	1
1	1	0

(b)

Figure 18.6 (a) Logic symbols for AND and NAND gates. The upper symbols are the recommended IEC/UK standard, the lower ones are the American standard. Both are widely used. (b) Truth tables for AND and NAND gates.

OR/NOR gates

Figure 18.7 shows the symbols and truth tables for OR and NOR gates. In an OR gate the output is 1 if **either** of the inputs is 1. In a NOR gate, the output is 0 if either of the inputs is 1.

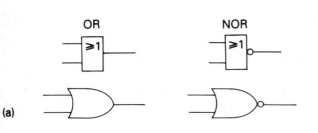

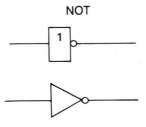

OR | | | NOR | |
| | | | | |

OR				NOR		
Inputs		Output		Inputs		Output
0	0	0		0	0	1
0	1	1		0	1	0
1	0	1		1	0	0
1	1	1		1	1	0

(b)

Figure 18.7 (a) Logic symbols for OR and NOR gates. (b) Truth tables for OR and NOR gates.

NOT gate

Perhaps the simplest of all, the NOT gate simply inverts its input, giving a 0 output for a 1 input and vice versa. It is sometimes called an **inverter** (Figure 18.8).

NOT

Figure 18.8 NOT gate: inverts its input.

EX-OR gate

Figure 18.9 gives the symbols and truth tables for an EX-OR, or **exclusive-OR** gate. The output is 1 when just **one** of the inputs is 1. When neither of the inputs is 1 or both of the inputs are 1 then the output is 0.

EX-OR

(a)

Inputs		Output
0	0	0
0	1	1
1	0	1
1	1	0

(b)

Figure 18.9 (a) Logic symbols for EX-OR gate. (b) Truth table for EX-OR gate.

Integrated circuits for digital logic

These gates are all available as integrated circuits. For example, Figure 18.10 shows an integrated circuit in a 14-pin DIL package, containing four NOR gates. This is a real IC, type SN74LS02. Digital logic circuits can be constructed from their logic diagrams very easily. The SN series of gates uses what is known as TTL (transistor–transistor logic) technology. It requires a stable 5 V power supply, and can work at high frequencies, into the low hundreds of MHz.

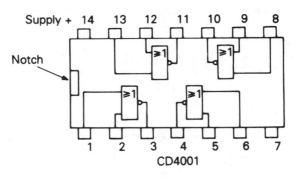

Figure 18.10 Four gates in a DIL IC package.

A different type of IC, known as **CMOS** (complementary metal oxide semiconductor) is easier for students to use. It is very tolerant of supply voltage (3–18 V), uses hardly any power, and so can be operated for many hours from a small 9 V battery. Its maximum operating frequency is lower, at up to around 5 MHz.

A bistable multivibrator made with logic gates

It is possible to make multivibrator (and other two-

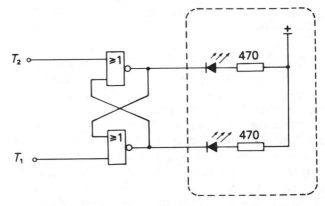

Figure 18.11 A bistable multivibrator built with NOR gates.

state) circuits with logic gates. Look back at the multivibrator circuit in Chapter 15, Figure 15.8 (did you build one?) and compare it with the logic gate equivalent in Figure 18.11.

The part of the circuit in the dotted box is to indicate, with LEDs, which side of the bistable is 'on', and is not essential to the circuit's working. Figure 18.12 shows a version you can build, using an integrated circuit with four NOR gates (two of them won't be used). The IC used, a CMOS CD4001, is cheap and easily obtainable, but is of a type that can be damaged by electrostatic voltages. But if you are careful and use a suitable soldering iron (or fit the IC into a socket) you should have no problems in constructing this circuit.

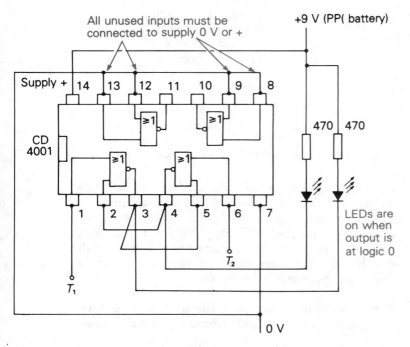

Figure 18.12 *A bistable multivibrator built with logic gates.

The circuit will work with a small 9 V dry battery as a power source. Momentarily connecting T_1 or T_2 to the positive supply will flip the circuit into one or the other stable state.

A bistable, by its nature, 'remembers' one of two conditions. It can retain – for as long as we want it to – a memory of which state it is in. So if, in Figure 18.12, we mark one of the LEDs with a '1' and the other with a '0', the circuit becomes a 'memory element' that can remember whether a single digit is either one or zero. And that is the basis of all digital computers.

Computers use a system of counting known as the **binary** number system. Instead of the ten numbers we are used to (the **denary** system), it has just two – zero and one:

denary	binary
0	0
1	1
2	10
3	11
4	100
5	101
6	110
7	111
8	1000
9	1001
10	1010
11	1011
12	1100
13	1101
14	1110

... and so on.

All the arithmetic operations that can be done with ordinary (denary) numbers can equally well be done with binary. All those ones and zeros look very confusing to the human eye, but of course they are ideal for electronic processing.

Binary arithmetic can be implemented in quite simple digital logic systems, for addition, multiplication, division and subtraction. Because these basic operations can be performed at tremendous speed, computers are capable of almost any mathematical calculation that a person can do, but much faster and with perfect accuracy.

An astable multivibrator made with logic gates
It is very easy to make an astable circuit using CMOS logic gates. A simple circuit is given in Figure 18.13; it can be made using a CMOS CD4010, which has six NOT gates in a 14-pin DIL package. The operating frequency of the circuit is approximately

$$f = 1.4\, RC$$

where f is in Hz, R is in megohms and C is in

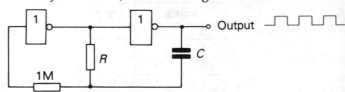

Figure 18.13 An astable multivibrator built with logic gates.

microfarads. (The formula is actually ohms and farads, but you can use megohms and microfarads more conveniently.) The output is a square wave.

A monostable multivibrator made with logic gates

A CMOS monostable is, if possible, even simpler: see Figure 18.14. With input T_1 at 1, the output is also 1. If the input is taken to 0, then the output goes to 0, but only for a time (t seconds) equal to

$$t = 0.5\,RC$$

after which it returns to 1. Given the information that CMOS logic changes state at half the supply voltage, you can easily work out for yourself how the circuit in Figure 18.14 works.

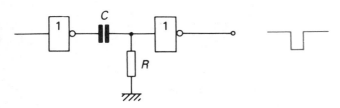

Figure 18.14 A monostable multivibrator built with logic gates.

Computer technology

Digital logic circuits form the basis of all digital computers, but you will seldom find simple gate ICs in a computer. Instead, literally millions of gates are made on a single integrated circuit, connected to form a complete – and often hugely complicated – system or subsystem.

Digital logic is important because it is the key to understanding computer technology, which these days is not confined to computers. Compact disc players, for example, are based on computer technology, as are telephone exchanges, television studio equipment, calculators and watches.

It may, at this stage in your studies, seem a long way from logic gates to computers, but having come this far you should now be equipped with enough knowledge to look at some of the modern

world's most complex, useful and fascinating machines.

■ CHECK YOUR UNDERSTANDING

● Digital logic gates have two possible input and output states: 1, usually corresponding to a high positive voltage, and 0, usually corresponding to a low (or zero) positive voltage.
● The most common types of gate are: NAND (negated-AND), NOR (negated-OR), EX-OR (exclusive-OR) and NOT (inverter). (Students! Make sure you know the truth tables for each kind!)
● Many different circuits and systems can be made with logic gates. The three types of multivibrator, for example, can be implemented very easily using standard logic gates.
● Computers use digital logic to perform calculations and manipulations of data.
● Electronic counting and arithmetic is carried out by using the **binary** system of counting, which has only two digits (1 and 0) and so can be represented by the two possible states of a logic gate.

REVISION EXERCISES AND QUESTIONS

1 Draw a truth table for the gate implementation given below:

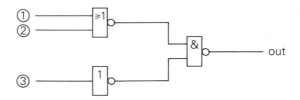

2 Draw the symbols for:
   i) two-input NAND gate;
   ii) three-input NOR gate;
   iii) two-input AND gate;
   iv) NOT gate.
3 What is the denary equivalent of binary 10000?
4 Why are bistable circuits important in computers?

Answers to questions and answering hints

Introduction

This section provides you with all the answers to the variety of questions and exercises given in the book. Always try a question or exercise yourself before you look at the answer. This will increase your understanding of the topic and give you practice in answering questions. If you are not sure of a particular answer, re-read the relevant section or chapter in the book to revise the work. You need to understand why a question has a particular answer, so that you can apply your understanding to similar types of question or exercise in your examinations and course assignments.

The book contains a variety of types of question and exercise. Find out the types of question that you will be expected to answer and their pattern. If possible, obtain past papers to support your work and revision. Some of the questions in the book require longer answers. We have provided hints on how to tackle these questions, and on the range of topics that you should include. Practise giving full answers to these questions and then check the answering hints to see that you have included all the relevant topics.

To revise a topic quickly you can also refer to the 'Check your understanding' sections given at the end of each chapter, and the list of key words with definitions given at the end of the book.

Hints to answering questions in examinations and course work

- Read all the questions carefully before you try anything. Make sure that you understand what each question is asking you to do.
- Plan the time that you will spend on each question. Use the marks as a guide: the more marks a question is worth, the more time it is worth spending on it.
- If you have a choice of questions, try to make your choice and stick to it. Don't change your mind halfway through the examination.
- Make sure that you earn all the 'easy' marks. Do not spend too long on a question you find difficult. Leave it; if you have time, you can try it again later when you have finished all the other questions.
- Keep an eye on the time. Make sure that you try all the questions you are required to answer.
- Always present your work as clearly as you can, whether you are writing or drawing. Make your work easy to follow for the examiner or assessor.
- Try and allow some time at the end to check your answers and improve them.
- In practical work, make sure that you understand what you are being asked to do by re-reading the question before you start. Follow all instructions carefully.

CHAPTER 1

1 The very first thing you must do if you see somebody who you believe is a victim of electric shock is to get them away from the source of electricity. **Without touching the victim**, you should therefore quickly decide whether you can turn off the power, or whether it is better to push the person clear with an insulator: something non-metallic that will not conduct electricity, such as a dry broom handle or branch. **Caution:** if the person is touching high-voltage power transmission lines, you should not try to move him, but should get help as quickly as you can, because a broomstick or something similar

will not provide you with enough protection.

2 Keep up artificial respiration for as long as it takes help to arrive. People have been known to recover even after several hours.

3 Electrical burns can be more severe than they look, so always treat them as very serious, even if they don't look very bad. Get a doctor to examine any such burns as soon as possible.

4 If you find yourself confronted by a potentially serious fire, you must **raise the alarm** before you do anything else. If an electrical appliance is on fire, you must next **disconnect the electricity** before you attempt to put it out.

5 No, car batteries can be dangerous because of the amount of current they can supply. Connecting the battery terminals together with anything that carries electricity can result in burns, a fire, or even an explosion. Also, a car battery is filled with strong acid, strong enough to burn your skin or clothes, or blind you if it gets in your eyes.

CHAPTER 2

1 The effects of an electric current are:
   i)   the heating effect;
   ii)   the lighting effect;
   iii) the chemical effect;
   iv) the magnetic effect;
   v)   the mechanical effect.
   All these effects are reversible.

2 The stylus (needle) of the pick-up is moved backwards and forwards by the wavy groove in the record. This bends the piezoelectric crystal inside the pick-up backwards and forwards. The **mechanical effect** of an electric current causes a p.d. to appear between opposite sides of the crystal. The amount of this p.d. varies in sympathy with the movements in the groove, and thus with the original sound recording.

3 i)   e.m.f. = electromotive force.
   ii)   p.d. = potential difference
   iii) PVC = polyvinyl chloride (a plastic commonly used to insulate wires)
   iv) µF = microfarad (one millionth of a farad, the unit of capacitance)
   v)   LED = light-emitting diode.

4 A fuse is fitted to disconnect the electricity supply to an appliance under fault conditions.

5 i) $V = IR$
   ii) $R = V/I$
   iii) Ohm's law calls for ohms, volts and amperes (not milli- or micro- anything!).

6 The resistance across $T_1 - T_2$ is 53 Ω.
Calculate the parallel resistors first:
   $3 \times 1/18$
   $3 \times 0.0555 = 0.166$
   $1/0.1666 = 6$
Add the series resistor:
   $6 + 47 = 53$

7 'The sum of the currents flowing into any junction in a circuit is always equal to the sum of the currents flowing away from it.'
'The sum of the potential difference in any closed loop of a circuit equals the sum of the electromotive forces in the loop.'

8 Size for size, electrolytic capacitors can have a much larger value of capacitance than other types of capacitor.

9 The current flowing is 6.25 A (1500/240 = 6.25). You should fit the next fuse **larger** than this current. In most countries (depending on their national standards) this would be either 10 A or 13 A.

CHAPTER 3

1 Direct current always flows in one direction only through a conductor or circuit; alternating current flows backwards and forwards, first in one direction and then in the other direction. The frequency of a.c. is measured in hertz (Hz). One hertz is equal to one cycle per second. The voltage is usually give as the rms (root mean square) voltage.

2 A primary cell produces an e.m.f. that can be used to power various items of electrical and electronic equipment. It cannot be recharged.

A secondary cell also provides an e.m.f., but can be recharged by applying a source of e.m.f. to it.

Examples of primary cells: Leclanché cell, manganese–alkaline cell, mercury cell, silver oxide cell, zinc–air cell, lithium cell, solar cell (although the solar cell doesn't use chemicals, of course). Secondary cells: lead–acid cell, nickel–cadmium cell, nickel–iron cell.

3 You will find that there are more and more questions that will need you to use your wits! The Leclanché cell has a p.d. of 1.5 V, which doesn't divide evenly into 4.8 V (the nearest you can get is 4.5 V). It is more likely to be a cell that has a p.d. of 1.2 V, which suggests that it might be a nickel–cadmium cell. The relatively high stated current capability of 3 A tends to confirm this identification.

4 The requirement for an electric current to be

induced in a wire placed in a magnetic field is that the field strength is changing. Either the wire could be moving, or the field itself could be changing in strength.

5 The rms (root mean square) value of an alternating current or voltage is the 'average' current or voltage, and is generally used as a measure of these quantities. The rms value of a sine-wave alternating current or voltage is 0.707 times the peak value.

6 Conventional current is current that is described as flowing from positive to negative potentials. It is the opposite of **electron current**, which is the flow of electrons from negative to positive potentials.

CHAPTER 4

1 Your sketch should have the main features of Figure 4.1.

2 Your sketch should have the main features of Figure 4.4.

3 Your essay should contain references to the following:

Atomic power: Potentially very dangerous; a serious accident could kill thousands of people and pollute hundreds of square kilometres of land. Waste products from fission reactors stay radioactive for thousands of years and are difficult to dispose of or store safely. On the other hand, atomic power does not pollute the atmosphere, river or sea and will never all be used up. Quite expensive.

Oil: Cheap, plentiful at the moment, not very dangerous. Will eventually (within tens rather than hundreds of years) all be used up. Causes serious pollution of the air, and of the sea if oil tankers sink. Should oil be conserved to make useful chemicals?

Hydroelectric power: Cheap once you have built the power station (and a dam if that is needed as well). Can only be sited where there are suitable natural resources (falling water). May cause environmental damage (flooding, or damage down-river) but otherwise completely pollution-free.

4 i) 1 horsepower = 746 watts.
ii) The unit of energy derived from the watt is the **joule**, or watt-second. The **kilowatt-hour** is commonly used for larger amounts of energy.

5 Size for size, a permanent-magnet motor is much more powerful than a shunt-wound motor.

Because the current it uses does not have to produce the stator field, it is more energy-efficient.

6 The electrical resistance of power cables is significant, and the higher the voltage used for long-distance distribution, the less power will be wasted as heat. Very high-voltage cables would require impossibly thick insulation if they were to be buried under the ground. Suspending the cables on overhead pylons and letting the surrounding air insulate them is the only practical solution to power distribution.

CHAPTER 5

1 Although the details depend on national regulations and standards, almost every domestic power installation will have the following fuses (starting at the point where the main power cable comes in from the power company):
i) main fuse (60–80 A);
ii) main switch (there may be a 50 A fuse in this, but not always);
iii) distribution fuseboard (various fuses up to 30 A);
iv) plug fuses (up to 15 A).

2 Your sketch should look like the circuit in Figure 5.3.

3 Advantages: Fluorescent lights produce more light for a given amount of power (they are actually at least four times more efficient than incandescent lamps). They last longer (up to eight times) and give out less heat. The diffuse light they produce is often more pleasant for shops, offices, etc.
Disadvantages: Fluorescent lamps are more expensive than incandescent lamps because of the additional components – ballast, starter – built into the fitting. When they are first switched on, they either come on dimly to start with or flash a few times before lighting properly.

4 Mains leads to electrical appliances in the UK:
live = brown
neutral = blue
earth = green/yellow stripes
House wiring:
live = red
neutral = black
earth = no sheath OR green

CHAPTER 6

1 Good moving-coil meters in common use

usually have an accuracy of ±5%, or even ±1%.

2 Advantages of a digital meter include:
   i)   easier to read
   ii) more robust
   iii) smaller
   iv) more accurate
   v)   often auto-scaling
   Disadvantages:
   i)   can't be used with continuously changing values
   ii) require a battery or other power source
   iii) the display sometimes seems more accurate than it is.

3 The oscilloscope will indicate (or in better instruments, measure) the **frequency, amplitude** and **shape** of a waveform.

4 A frequency generator produces a signal of known frequency, amplitude and waveform. Frequency generators can be obtained that will produce signals from low audio frequencies to the highest radio frequencies. Modulation of radio-frequency signals is often an option. The signal can then be traced through a circuit or system with a suitable oscilloscope.

5 Two measures of sensitivity are:
   i)   full-scale deflection (FSD), which is the amount of current required to move the meter needle to the end of its scale;
   ii) sensitivity in ohms per volt, which is the resistance of the meter coil divided by the voltage needed to move the meter needle to the end of its scale.

CHAPTER 7

1 Your drawing should look more or less like Figure 7.1. The important points to include are the electromagnet, and a moving armature that opens one contact and closes another when the electromagnet is energised.

2 A speaker consists of a paper, composition, or plastic cone mounted so that it can move to and fro. The narrow end of the cone has a coil wound round it, and is positioned in the field of a powerful permanent magnet. An electrical signal passing through the coil will move the cone in or out (because the magnetic effect of a current induces a magnetic field in the coil which is either attracted to or repelled by the field of the permanent magnet). If an audio-frequency electrical signal is applied to the coil, the speaker cone will move in sympathy with the signal and

will produce a corresponding sound by moving the air round the cone.

3

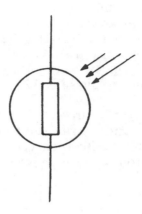

A light-dependent resistor changes its resistance according to the amount of light falling on it. The brighter the light, the lower the resistance. (Extra marks if you have discovered that the effect is substantial and LDRs can change from several megohms (dark) to less than a kilohm in bright light.)

4 Moving-coil microphones are used in telephones in preference to electret microphones because they need a power source, and in preference to crystal microphones because they have a high-impedance output that is unsuitable for use with transistor circuits. (Note that the sound quality is not important here because the telephone system amplifiers are much more of a limiting factor than the microphone.)

CHAPTER 8

1 i) An **intrinsic** semiconductor is a material that naturally has properties intermediate between those of a conductor and those of an insulator.
   ii) An **extrinsic** semiconductor is a material that has been given semiconducting characteristics artificially, by adding minute traces of other elements.

2 i) A ***p*-type semiconductor** is a semiconductor that has extra (positive) holes – 'missing' electrons – in its structure.
   ii) An ***n*-type semiconductor** is a semiconductor that has extra (negative) electrons in its structure.

3 Only the **conduction band,** the outermost energy band (which may sometimes be empty), and the **valence band,** the next band below the conduction band, are important in electronic interactions.

4 In order to move out to the next higher energy band, an electron has to gain energy.

CHAPTER 9

1

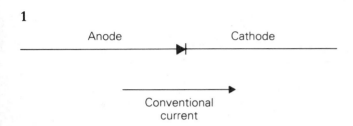

2 The key to this is remembering that the silicon diode has a forward voltage drop of about 0.7 V, unless it is hot. The power dissipated can therefore be calculated by the usual formula, $P = V_f I$.

$P = 2 \times 0.7$ W

$P = 1.4$ W

3

(Extra marks if you used a preferred value of resistance! If you used a resistor of more than 300 Ω you probably forgot to allow for the voltage drop of the LED.)

4

5 i) A **photodiode** is a form of *pn* diode in which the reverse leakage current changes according to the amount of light falling on the junction: the more light the higher the current.

ii) A **varicap diode** is a diode that makes use of the junction capacitance of a reverse-biased diode. Changes in applied voltage alter the junction capacitance.

6 i)   The LED will light.

ii)   The LED will light. (Trace the path through the diodes for either direction of applied voltage.)

iii) The LED will not light. (The reason is that the sum of the forward voltage drop of two silicon diodes and the LED – about 3.5 V – exceeds the supply voltage and no current will flow.)

CHAPTER 10

1

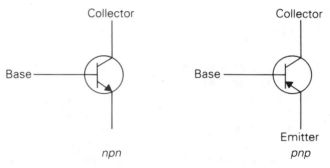

2 i)   The **large signal current gain** of a transistor is the collector current divided by the base current: the greatest amount that the transistor can amplify a signal.

ii)   A **heat sink** is a piece of metal that is fixed to a transistor – usually a power transistor – to conduct heat away from the transistor junction to prevent it overheating.

iii)   The **leakage current** (unless quoted more specifically, e.g. 'base leakage current') is the current that flows through the emitter–collector junction with the base terminal disconnected. It depends on the applied voltage and on the ambient temperature.

3 A transistor is a three-terminal device in which a small current flowing between the transistor's base and emitter terminals allows a larger – and proportional – current to flow between the emitter and collector terminals.

4 The **emitter current** is the current flowing through the emitter at any given instant. It is always equal to the sum of the base and collector currents.

CHAPTER 11

1

2 In a transistor the current flowing between the collector and emitter is proportional to the base current. The FET equivalent, the current flowing between the source and drain, is controlled by the **voltage** on the gate, rather than by current.

3 FETs are damaged by electrostatic voltages, so it is vital to take precautions to avoid static electricity. Even handling FETs can destroy them in some circumstances.

4 '... integrated circuits.'

CHAPTER 12

1 Silicon is used as the basis for integrated circuits. Its advantages are as follows:
   i) It is very cheap.
   ii) Many different classes of semiconductor device can be made with silicon.
   iii) Silicon dioxide can easily be formed on its surface, and silicon dioxide is a good insulator, chemically very resistant, and prevents diffusion.

2

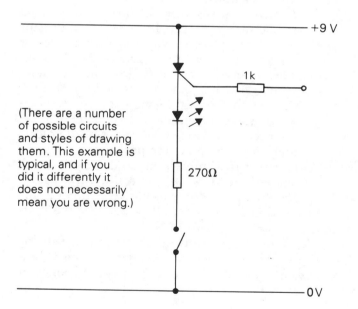

(There are a number of possible circuits and styles of drawing them. This example is typical, and if you did it differently it does not necessarily mean you are wrong.)

3 A **triac** is a bidirectional thyristor. It has two anodes and a gate. A small current flow in either direction between the gate and one of the anodes will trigger conduction of a much larger current – also in either direction – between the two anodes. This current will continue to flow until it drops to a value near zero, when the triac will switch off (if there is no gate current flowing).

CHAPTER 13

1

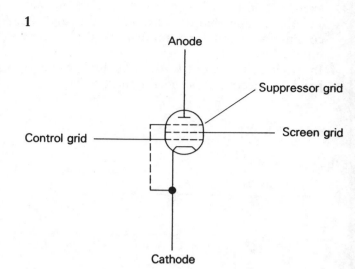

2 The heater heats the cathode up to a temperature of about 1750 °C, at which temperature the cathode – specially treated with thorium – emits electrons.

3 Valves are not used very much these days, but are still found in large radio and television transmitters.

4 A triode valve is a three-terminal vacuum-state device in which a voltage applied to the grid controls a proportional current flowing between the anode and cathode.

CHAPTER 14

1

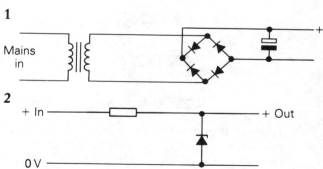

2

Or better than this is:

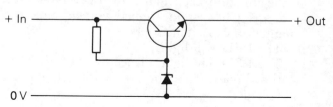

3

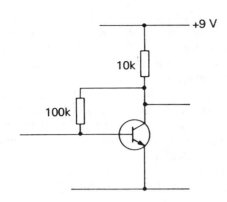

4 Capacitor coupling allows a.c. signals to be passed from the output of one stage to the input of another without upsetting the bias conditions of either stage.

5

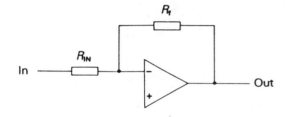

6 The main disadvantage of Class A amplifiers is their large power consumption even when there is no output. This causes a consequent problem of large heat output.

7 Class B amplifiers divide the signal into two halves, having two output transistors, one to deal with the positive excursions of the signal, the other to deal with the negative excursions. Where there is no signal, neither transistor is conducting very strongly and there is little power used. (The power consumption of a Class B amplifier is roughly in proportion to the loudness of the output.)

8 The **input impedance** is the impedance (resistance to an alternating current) that is 'seen' by a signal source feeding a signal into an amplifier.

CHAPTER 15

1 An oscillator is a circuit that generates a continuously changing voltage or current, usually (but not necessarily) at a fixed frequency.

2 $f = 1/2\pi\sqrt{LC}$

3

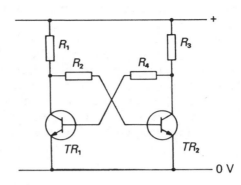

If either transistor is forced to turn on or off – by connecting its base to its emitter or by applying a current to the base through a resistor – the circuit will remain with that transistor either on or off indefinitely.

4 The astable multivibrator circuit produces a **square wave** output.

CHAPTER 16

1 $\lambda = v/f$ (v = the speed of light in m/s, f = the frequency in Hz)
$\lambda = 300\ 000\ 000/12\ 500\ 000$
$\lambda = 24$ m

2 i) **Amplitude modulation** adds information – usually audio – to a carrier wave by altering the amplitude of the waveform according to the modulating waveform.

ii) **Frequency modulation** adds information to a carrier wave by slightly varying its frequency according to the modulating waveform.

iii) AM was discovered and used first because it is very simple to detect an AM waveform, using only a diode as a demodulator.

3 If a receiver is built with a series of fixed-frequency tuned amplifier stages to amplify the signal, it makes the receiver very selective. The superheterodyne system enables such a receiver to be tuned to different frequencies by altering only the tuning of the radio-frequency stage, along with that of the local oscillator, to keep the intermediate frequency the same.

4 **Interlaced scanning** reduces the flicker on the picture by apparently doubling the rate at which the screen is scanned from 50 or 60 Hz to 100 or 120 Hz.

5 The colour television picture tube has three electron beams, one for each of the three primary colours: red, green and blue. By additive mixing,

these three can produce any other colour, including white.

6 The answer to this question is a practical one. Standardising on one intermediate frequency (or on a very few) enables component manufacturers to produce a range of intermediate frequency transformers (IFTs) that are pretuned to the right frequency, with the expectation that many different radio manufacturing companies will want to buy them.

CHAPTER 17

1 Any component that has multiple connections and has to be removed from a PCB should be released with a **solder sucker**. Remove the solder from each terminal and make sure that the lead is free by pushing it to and fro. Use the soldering iron to ease it if necessary. When all the terminals are free, just lift the relay out.

2 The clue is the fact that the transistor is getting hot. The 3.3 kΩ collector resistor should prevent it from dissipating more than a few milliwatts. If it is getting hot to the touch, then either the 3.3 kΩ resistor has failed (very low resistance or short-circuited) or the 47 kΩ base bias resistor has done the same. There is a small chance that *both* the 10 kΩ collector resistor in the preceding stage **and** the 10 μF coupling capacitor have short-circuited. Alternatively, the PCB may be shorted.

3 Try a new battery in it. More than a fifth of all 'repairs' to domestic equipment are something silly like this. If it isn't the battery, use a signal generator and an oscilloscope to trace the faulty stage. Failed output (speaker) coupling capacitors give these sort of symptoms.

4 The symptoms are very characteristic of a build-up of dirt on the sound head. Clean the head with a cloth and a little alcohol, and if that doesn't work, try re-aligning it. Remember that 'mechanical' failures like this are more common than electronic faults.

5 If the transformer is getting hot it is probably because it is trying to deliver too much current. Where might wasted power be going? One possibility is that it is being dissipated in that big smoothing capacitor. Electrolytic capacitors can fail in such a way that they develop a very large leakage current, and behave as if they had

a low-value resistor in parallel with them. Try changing it. Otherwise disconnect the power supply from the circuits it is powering and see if it still gets hot. If it does, it's probably an insulation failure inside the transformer, and you'll need a replacement.

6 If somebody asks you to try to repair a personal computer, give it back. Not even the professionals try to mend them. IBM PC parts are mostly interchangeable, so the machine should be taken to a dealer who can advise about replacement, either of the 'motherboard' and power supply, or of the whole thing.

CHAPTER 18

1

1	2	3	Out
0	0	0	0
0	0	1	1
0	1	0	1
0	1	1	1
1	0	0	1
1	0	1	1
1	1	0	1
1	1	1	1

2

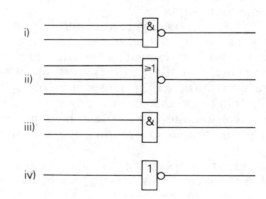

3 10000 binary = 16 denary
4 Bistable circuits can 'remember' one of two states, which can correspond to a '1' or a '0'. Thus – to use the example from the previous question – the binary number for 16 (denary) can be stored in five bistable circuits.

Key words and definitions

A Ampere (amp); the unit of electric current.

a.c. Alternating current.

AF Audio frequency.

alternating current An electric current that alternates in its direction of flow. The frequency of alternation is given in hertz. *Compare* **direct current**.

AM Amplitude modulation.

amplifier An electronic system that increases the power of a signal.

amplitude modulation A system of modulating a carrier in which the amplitude of the carrier is changed in sympathy with the modulating signal.

analogue A system in which changing values are represented by a continuously variable electrical signal.

armature In any machine involving magnetism, the armature is the moving part. Examples are the rotating part of an electric motor, and the moving part of a moving-iron meter.

astable A circuit which has no stable condition, and oscillates at a frequency determined by circuit values. *See also* **oscillator**.

audio Relating to a system concerned with frequencies within the range of human hearing.

base One terminal of a bipolar transistor. A small controlling current flows between the base and the emitter.

binary A number system to the base 2.

bipolar transistor A transistor in which current is carried through the semiconductor both by holes and by electrons.

bistable A system which can have two stable states, and which can remain in either state indefinitely.

breakdown A sudden loss of insulation properties, often resulting in a rapid and large current flow. Typically, breakdown might occur in a semiconductor device operated at too high a voltage.

capacitor A component used in electronic circuits, exhibiting the property of capacitance.

CCD Charge-coupled device. A complex development of integrated circuits, used in television cameras for imaging.

chrominance In a television system, the part of the television signal concerned with colour.

Class A amplifier An amplifier in which the output transistor is operated at approximately half the supply voltage, resulting in a continuous heavy current flow, but potentially low distortion.

Class B amplifier An amplifier in which the output is shared between two transistors, resulting in much more efficient operation but potential problems from crossover distortion.

commutator The part of an electric motor or generator that conducts current to the armature windings, switching the current as the armature rotates.

conductor A material through which an electric current can flow relatively easily.

conduit A metal or plastic pipe for protecting electric cables from physical damage.

collector One terminal of a transistor. The current flowing between the collector and the emitter is controlled by the base current.

conventional current Electric current, regarded as flowing from positive to negative.

CRT Cathode ray tube.

crystal Usually refers to quartz crystal, used as a precision timing element in many circuits. May refer to a piezoelectric crystal pick-up, or (old usage) to a diode used for demodulation in AM radio receivers.

dB Decibel: one tenth of a bel, the unit of relative power. *See* **decibel**.

d.c. Direct current.

decibel One tenth of a bel. A measure of power, based on a logarithmic scale. Symbol dB. The decibel is a convenient unit for representing a very large range of powers.

demodulation The recovery of a modulating signal from a modulated carrier.

denary The 'normal' number system, to the base 10.

diac A bidirectional breakover diode. Often used for triggering a triac.

digital logic An electronic system implemented with digital logic gates.

digital electronics The branch of electronics concerned with the processing of digital systems, usually in binary.

DIL-pack The standard package used for integrated circuits.

diode A component, either semiconductor or thermionic, that permits current to flow through it in one direction only.

direct current An electric current that flows steadily in one direction. *Compare* **alternating current**.

discrete Used to refer to systems constructed from individual components – such as transistors, capacitors, diodes and resistors – as opposed to systems made using integrated circuits.

dissipate (usually of heat) Get rid of – usually by conduction or radiation.

doping The addition of tiny amounts of impurities to semiconductor material during the manufacture of semiconductor devices.

electromagnetic Involving the use of electrically induced magnetism. Many important electrical devices involve this principle: for example, generators, motors, relays and speakers.

electroplating The process of depositing a layer of metal on a conductive base (usually also metal) by means of electricity flowing through a solution of a metallic salt.

electrolysis Conduction of electric current accompanied by the transfer of matter, resulting in chemical changes at the electrodes.

e.m.f. Electromotive force. The force that tends to cause movement of electric current around a circuit.

emitter (1) One terminal of a bipolar transistor. The emitter is common to the base and collector circuits.
(2) (thermionics) A substance or device that emits electrons.

energy The capacity for doing work. Energy is usually measured in joules or kilowatt-hours.

extrinsic semiconductor A semiconductor material produced artificially by the addition of impurities.

F Farad: the unit of capacitance. *See* **farad**.

farad Unit of capacitance. The farad is a very large unit, the largest practical unit being the microfarad.

ferrite A finely divided ferrous dust, suspended in a plastic material. Ferrite has useful magnetic properties, but does not conduct electricity.

field timebase In television, the oscillator used to control the vertical scanning of the picture.

field-effect transistor A type of transistor characterised by a very high input resistance.

fire blanket A blanket made from some fireproof material, used to smother small fires.

fluorescent lamp An electric lamp in which light is produced by making a phosphor coating fluoresce (glow).

flux (literally, a flow) Various meanings, but usually a resin added to solder in order to prevent the formation of oxides on the material being soldered. Also used to talk about a hypothetical 'flow' of magnetism, hence 'magnetic flux'.

FM Frequency modulation.

frequency The number of waves, vibrations or cycles of any periodic phenomenon, per second. Unit is the hertz.

gain The factor by which the output of a system exceeds the input.

gate (1) A component in digital logic circuits
(2) One terminal of a field-effect transistor, or other semiconductor device (usually the controlling terminal).

Ge Chemical symbol for germanium, a semiconductor.

H Henry. *See* **henry**.

Hall effect A change in the way that current flows through a conductor or semiconductor when subjected to a magnetic field.

henry Unit of inductance. Symbol H.

hertz The unit of frequency. One hertz equals one cycle per second.

Hz Hertz. *See* **hertz**.

IGFET Insulated gate field-effect transistor.

impedance The ratio of the voltage applied to a circuit to the current flowing in the circuit. Similar to resistance, but applicable to alternating currents and voltages.

incandescent lamp An electric lamp in which light is produced as a result of heating up a filament until it is white-hot.

induction motor An electric motor in which there are no electrical connections to the armature, current being induced in the armature windings magnetically.

inductor A component exhibiting a known amount of inductance.

inductance (self-induction) a property of an electrical circuit or component such that it resists any change in the current flowing through it. (Usually such components are electromagnetic.)

insulator A material through which electric current will not easily flow.

integrated circuit An electronic system, or part of a system, produced on a silicon chip using microelectronic techniques.

intermediate frequency In radio and television, the frequency generated as a result of mixing the local oscillator and incoming signal.

joule Unit of energy; the watt-second. The equivalent of a power of one watt, maintained for one second. *See* **kilowatt-hour**.

JUGFET Junction gate field-effect transistor.

kilowatt-hour Measure of energy, used in dealing with relatively large amounts of energy, such as the consumption of mains electricity.

LC oscillator An oscillator using an inductor and a capacitor in a resonant circuit as a timing element.

LED *See* **light-emitting diode**.

light-emitting diode (LED) An electronic component in which electric current is converted directly into visible or infrared light.

line timebase In a television, the oscillator circuit concerned with horizontal scanning of the picture.

logic Usually used as an abbreviation for 'digital logic', referring to systems involving logic gates.

luminance In television, the part of the signal concerned with the brightness of the image on the tube.

microfarad one thousandth of a farad. The largest practical unit of capacitance used in electronics.

modulation Variation of the frequency, phase or magnitude of a high-frequency waveform in accordance with a waveform of lower frequency.

monostable A system with a single stable state.

MOS Metal oxide semiconductor.

MOSFET Metal oxide semiconductor field-effect transistor.

multimeter   A general-purpose measuring instrument, usually able to measure resistance, current and voltage.

negative feedback   Feedback applied to a system in such a way that it tends to reduce the input signal that results in the feedback. *See* **positive feedback**.

npn   Negative–positive–negative (although always pronounced 'en-pea-en'); refers to one of the two alternative types of bipolar transistor.

NTSC   National Television Standards Committee. The American body that defined the American television standard. 'NTSC' is used to refer to the type of TV system used in the USA.

operational amplifier   A highly stable, high gain, d.c. amplifier, usually produced as a single integrated circuit.

oscillator   An electronic system that produces a regular periodic output.

oscilloscope   An instrument for displaying electrical waveforms on a cathode ray tube.

Ω   Ohm: the unit of resistance.

PAL   Phase alternation by line. The colour television system used in the uK and many other countries. It has advantages over the NTSC system that preceded it.

passive component   A component that does not involve the control of electrons in a thermionic or semiconductor device.

PCB   Printed circuit board.

PD   Potential difference. The difference in electrical states existing between two points.

pentode   A valve that is an improvement on the tetrode and has an additional grid between the screen grid and the anode, to prevent current flowing between the anode and screen grid.

photoresistor   Also known as an LDR (light-dependent resistor). A resistor whose value depends upon the amount of light falling on it.

piezoelectric effect   The direct conversion of electrical to mechanical energy, or vice versa, in some crystalline materials.

pnp   positive–negative–positive (although always pronounced 'pea-en-pea'); refers to one of the two alternative types of bipolar transistor.

positive feedback   Feedback applied to a system in such a way that the feedback tends to increase the input signal causing the feedback. *See* **negative feedback**.

potentiometer   A variable resistor having connections to each end of the track and also to the brush.

power   The rate of doing work. Power is usually measured in watts.

primary cell   A device that produces electrical energy, usually from chemicals. The chemical reactions are not reversible, and a primary cell cannot be recharged.

PVC   polyvinyl chloride. A tough plastic often used for electrical insulation.

quartz crystal oscillator   A very stable oscillator, depending for its stability on the electromechanical properties of a quartz crystal.

raster   The pattern of horizontal lines produced on a television screen.

relay   An electromechanical device in which an electric current closes a switch.

resistance   The property of a material that resists the flow of electrical current. Unit ohms.

resistor   A component exhibiting a known amount of resistance.

RF   Radio frequency.

rotor   The rotating part of an electrical generator.

secondary cell   A device that produces electrical energy from chemicals. The chemical reactions in a secondary cell are reversible, so the cell can be recharged by the application of a source of e.m.f. to the electrodes.

semiconductor   A material with properties that lie between those of insulators and conductors. Extensively used in modern electronics.

Si   Chemical symbol for silicon, a semiconductor.

SI units   The International System of units (Système International d'Unités), an agreed standard used throughout the world, with very few exceptions.

slip-ring   Part of an electric motor or generator, designed to conduct electric current to the rotating part of the machine. Unlike a commutator, a slip-ring does not switch the current.

solder   Alloy of tin and lead used for fixing and connecting metal wires and electronic components.

solder-sucker   A small hand-held tool for removing solder from a component or printed circuit. Used in conjunction with a soldering iron.

soldering iron   A hand-held tool for melting solder when building or repairing electronic or electrical circuits.

speaker   (loudspeaker) An electromechanical device for converting electrical energy into sound.

static electricity   electric charged developed in an insulating material. Static electricity is usually associated with high or very light voltages and exceedingly small currents.

superheterodyne   A radio receiver system in which the radio-frequency input is mixed with a frequency generated within the receiver to produce an intermediate frequency.

surface mounting   A technique of soldering tiny electronic components directly onto the surface of a printed circuit board. Enables boards to be made very small, but renders them very difficult to repair.

tetrode   A thermionic valve that is an improvement on the triode, having an extra grid – the screen grid – to allow increased gain and better high-frequency operation to be obtained.

thermionic   Electronic device involving electrons generated by heat, usually in a vacuum.

thyristor   A component similar to a semiconductor diode but having in addition a gate connection by which the component, normally non-conducting, can be triggered into conduction.

tolerance   Generally the amount by which a specified component value can vary from the marked value. Usually expressed as a percentage.

transistor   The basic semiconductor three-terminal amplifying device. A small current flowing between its base and emitter terminals controls a larger current flowing between the collector and emitter terminals.

transformer   An electromagnetic device consisting of two or more insulated coils, usually wound over a ferromagnetic core. Used for increasing or decreasing voltages and currents in a.c. circuits and systems.

triac A semiconductor component similar to the thyristor but which will conduct in either direction.

triode A device having three terminals. Usually used to refer to the triode valve, the simplest thermionic amplifier.

unijunction transistor A semiconductor device used in some oscillators.

valve (American 'tube') A thermionic device used for rectifying and amplifying. Now almost entirely superseded by solid-state devices except in very high-power applications such as broadcast radio transmitters.

V Volt: the unit of electrical potential.

varicap diode A semiconductor diode in which the junction capacitance varies according to the applied voltage. This effect is inherent in all semiconductor diodes, but in the varicap diode the property is deliberately enhanced. Used in tuning circuits in radio and television.

video (1) In television, the demodulated vision signal. (2) More generally, anything relating to the recording, replaying, transmission or reception of pictures.

wavelength The physical distance between two similar and successive points on an alternating wave.

Zener diode A semiconductor diode, used for voltage regulation. When the Zener diode is reverse-biased, it exhibits a sudden increase in conductivity at a certain specific voltage.

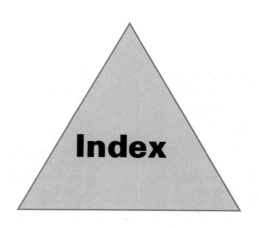

Index